Jet Engine Maintenance Techniques

Complete training data for basic gas turbine engine maintenance learning.

Syed Rumman Akhtar

Aircraft Maintenance Engineer

Preface & Acknowledgement

The book is designed to aid the students and learners in their day to day study. The chapters in this book discussed are about Jet Engine Maintenance.

This volume gives the information about the requirements of Aircraft engine Maintenance and contains safety precautions, basic procedures, locations and functioning of components. Since the maintenance of Aircraft Engine is most important and critical, all the materials connected with Aircraft engine Servicing and maintenance has been taken care as per EASA module 15 and covered up in this book.

This book will also be of great help to the students appearing for DGCA, AME knowledge examination.

I am highly thankful to my family for their guidance, encouragement and active involvement to make this book take its shape.

I would very much appreciate criticism, suggestion and detection of errors from the readers which will be gratefully acknowledged.

Best regards,

Syed Rumman Akhtar

Index

| | Composite Propellers, Blade Cuff Inspection, Governors, Balancing , Static Balancing, Out-of-Balance Repairs, Dynamic Balance, Checking Blade Angle, Blade Angle Adjustments, Ground-Adjustable Propeller, Counterweight, Hartzell Constant Speed Propellers, Hartzell Constant Speed Propellers, Blade Tracking, Troubleshooting, Hamilton Standard Hydromatic, Propeller Installation, Flanged Shaft, Fixed-Pitch Propellers, Constant-Speed Propellers, Turbopropellers, Splined Shaft, Propeller Safetying, Fuel Nozzle Testing, Testing PT6A Fuel Nozzles, Component Maintenance, Compressor Blades, Turbine Nozzles and Vanes, Inspection (Borescope) and their serviceability limits; Turbine Blade, Repairs for Turbine Nozzles, Vanes, and Blades, | |

## Chapter - 1

INSPECTION

AND MAINTENANCE

LUBRICATION SYSTEM

Internal lubrication of the engine is vital importance and the oil quantity should be checked daily or prior to each flight. Oil should be changed regularly by draining the sump and tank and refilling the system with new oil to the correct specification. oil screens should be cleaned and filter elements changed at the specified intervals but on removal should be inspected for the presence of metal particles which would indicate internal failure of the engine.

The oil cooler air passages should be checked for blockage and cleaned as necessary and all the parts of the oil system should be checked for cracks, security, chafing, leaks and damage during routine inspections.

During filter inspections, if metal particles noticed, the cause for metal presence is to be investigated. If small pieces of metal noticed drain the oil completely from the tank, run the engine until the oil temperature is stabilized. Drain the oil from the tank after stopping the engine, check the metal particles for the engine. If No traces of metal particles noticed in the oil system, then make the engine serviceable.

If metal particles found in the filter, dismantle the engine completely and find the source of metal failure If the failure of bearing noticed, replace the bearing. All other bearings should inspected and flushed thoroughly to remove traces of metal. All oil system components such as oil tank, oil pump scavenge pump, oil cooler, filters, oil lines should be inspected by dismantling and thoroughly flushed before assembly of the components. After assembly of the units, the units should be bench checked.

Before installing oil pipe lines in the oil system, the same is to be pressure tested.

The oil system, components including packing should be inspected for shelf life. Some of the oil system components are having overhaul life. There should be method of monitoring the life of the component and life reached to specified limits, the same to be overhauled.

During installations of oil system components to the engine, required installation procedures is to be followed as per overhaul manual.

The gasket and rings to be ensured for having a shelf life, and lubricated properly with approved lubricant during installations. After installation of components, check for oil leaks after a brief ground run.

While rectification of oil system snags, it is advisable to have oil flow diagram to trace the oil system problem.

Most turbine engines have a closed cycle oil system similar to that used on piston engines. Rotor main bearings and all accessory drives and gears are pressure lubricated, scavenge oil being either drained in to a wet sump or returned to a tank mounted on the engine casing.

The main compressor and turbine bearings on some engines are lubricated by a waste system which employs micro-pumps to provide a metered supply of oil to each bearing. The oil passes through the bearing and exhausts to atmosphere with the burnt gases. In this type of system both maximum and minimum permitted oil consumption figures are quoted.

Routine maintenance of the oil system consists of checking the level of oil in the oil tank and topping up as necessary with the appropriate type and grade of oil. When changing from one type of oil to another it may be necessary for the tank to be drained, partially refilled with the new oil and the engine run for approximately 15 minutes, exercising all oil operated controls as appropriate. The system should then be drained once more and refilled.

NOTE: Some synthetic oils used in turbine engines contain tri ortho cresyl phosphate or other additives which are highly toxic and should not be allowed to come into contact with the skin. It is recommended that suitable gloves should be worn by personnel continuously handling these oils.

Filters are provided in the oil system and should be removed for examination at -1-le intervals specified in the Maintenance Schedule. Light metal swat-if may be expected in new engines but a heavy deposit indicates a failure in the engine. Metal filters should washed in kerosene and dried with compressed air before refitting, but 'throw-away' type -paper filters should be renewed. It is also essential that the filter casing is flushed out to remove any residual contaminant.

Chip detectors and magnetic plugs are often fitted in the oil system and should be removed for examination at regular intervals. Some chit detectors can be examined in situ by the use of a 250 volt megger, with zero resistance between the centre of the detector and the casing indicating that sufficient metal particles are present to warrant removal and examination of the detector. This principle has been further extended so that detectors can provide an electrical signal for cockpit indication of particle build up, enabling in-flight monitoring and also automatic recording of necessary post-flight maintenance action.

LUBRICATION SYSTEM FAULTS AND OIL CONTAMINATION

Mechanical faults are sometimes detected by performance monitoring systems including oil pressure gauges, oil temperature gauges, metallic chip detectors, and other components. In addition, internal engine faults are detected during oil filter inspections and spectrometric oil analysis of oil samples. Typical lubrication system faults and oil contamination considerations include the following:

1.	Insufficient oil quantity - Follow the engine manufacturer's instructions for checking the oil level. In most cases, the oil level must be checked within a specified time after the engine has been operated.

2.	Check the indicating system to verify proper pressure gauge indications.

3.	Check the oil filter and screens for obvious signs of contamination or debris.

4.	Inspect external components of the lubrication system such as oil pump, pad to fuel heat exchangers for signs of cracks, FCOC, oil scavenge pump, gear box moulting and oil lines for sign of oil leakage.

5.	Inspect the oil pressure relief valve for proper functioning and attempt to adjust the oil pressure to specified levels.

6.	Check the oil pressure indicating system for proper operation.

7.	Check the oil pressure relief valve for proper operation and attempt to adjust the valve to achieve specified limits.

8.	Check the oil level. Insufficient or excessive oil quantity may cause fluctuating oil pressures.

9.	Check the oil pressure indicating system for proper operation.

10.	Check the airframe oil cooler system for leaks or blockage.

11.	Inspect and clean the oil filter and screens, as required.

12.	Check the oil level. Low quantities cause elevated lubrication system operating temperatures.

13.	Check the oil temperature indicating system for proper operation.

14.	Inspect the oil cooler for signs of leakage and proper thermostatically controlled bypass valve operations.

15.	Check the operation of the oil pressure relief valve.

16.	Consult the engine manufacturers troubleshooting instructions to isolate possible internal engine carbon or labyrinth seal failure.

17.	Check for leakage or blockage in the pressure and scavenge oil tubes and ports. Leaking scavenge and pressure ports and tubes may also cause fluctuating oil pressures or total loss of oil pressure due to the lubrication pump(s) losing their prime.

18.	Check for oil leaks in the oil filter element housing, the fuel heater, and the airframe oil cooler assembly.

19.	Inspect the tailpipe and around case seals for signs of oil. If leakage is evident, refer to the engine manufacturer's instructions for further guidance to isolate the source of the leak.

20. Perform an operational check of the chip detector by removing the sensing element. Once removed, inspect for metallic particles between the sensing elements. If the detector is found clean, repair or replace the detector.

21. If metallic particles are found, refer to the engine manufacturer's troubleshooting procedures and maintenance manual for information on how to detect the type of metal and probable sources.

EVIDENCE OF METALLIC PARTICLES IN THE ENGINE OIL AND OR FILTER

Perform a spectrometric oil analysis to determine the exact type of metal and quantity. Compare the results of the analysis to the engine manufacturer's guidance material to determine if the engine can continue in service, and the conditions that mu.si. be met for continued service, or it the engine must be removed for disassembly and repair. Spectrometric Oil Analysis Another technique used to detect hidden problems in turbine engines is done by performing a spectrometric oil analysis. A spectrometric oil analysis program, or SOAP, is available to aircraft operators to help detect developing problems in an engine. Spectrometric analysis for metal particles suspended in oil is possible because metallic ions emit characteristic light spectra when vaporized by an electric arc. Each metal produces a unique spectrum, allowing easy identification of the metals present in an oil sample. The wavelength of spectral lines identifies each metal and the intensity of the line is used to measure the quantity of that metal in a sample. 111 When participating in a spectrometric oil analysis program, periodic samples of oil are taken from the engine after shutdown or prior to servicing. Samples are taken from a sediment free location in the main oil tank and sent to an oil analysis laboratory. In the lab, a filrit of the used oil sample is picked up on the rim of a rotating, high purity, graphite disk electrode. A precisely controlled, high voltage, AC; spark is discharged from a vertical electrode to the rotating disk. When this occurs, the film of oil on the disk begins to burn. Light emitted by the burning oil passes through a series of slits, precisely positioned to detect the wavelengths of various metals. As light passes through the slit, photo multiplier tubes electronically convert the light waves into energy, which automatically prints the analytical results on the laboratory record sheets. The wear metals present are so small that they flow freely through an engine's system filters. The spectrometric therefore measures the particles that move in suspension in the oil and are too small to appear on either the oil screen or chip detector.

Alloyed metals in turbine engines may contain amounts of aluminium, iron, chromium, silver, copper, tin, magnesium, lead, nickel, or titanium. Silver is accurately measured in concentrations down to one-half part silver in one million parts of oil. Most other metals are measured accurately. in concentrations down to two or three parts per million. The maximum amount of normal wear has been determined for each metal of the particular system in the program. This amount is called its threshold limit of contamination and is measured by weight in parts per million (PPM). If after interpreting the results, the lab identifies a sharp increase of abnormal concentrations of metal, the lab will immediately notify.

The types of metals identified during a spectrometric oil analysis provide invaluable information to help you determine the source of the contamination. Engine manufacturers provide a list of engine components and the materials used in their construction. Consult this information to identify areas where to begin troubleshooting and for corrective actions to take when the source of contamination has been identified.

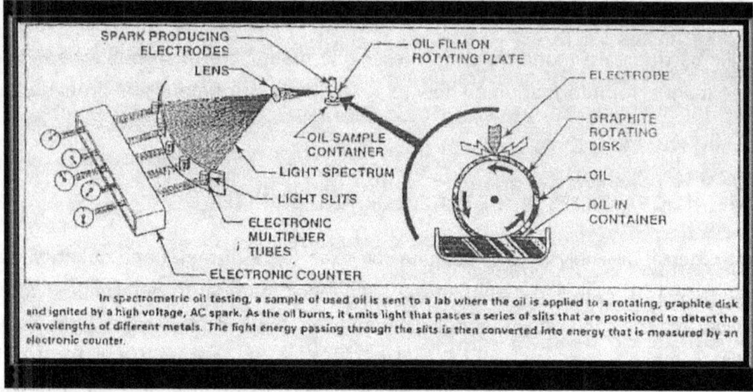

In spectrometric oil testing, a sample of used oil is sent to a lab where the oil is applied to a rotating, graphite disk and ignited by a high voltage, AC spark. As the oil burns, it emits light that passes a series of slits that are positioned to detect the wavelengths of different metals. The light energy passing through the slits is then converted into energy that is measured by an electronic counter.

LUBRICATION SYSTEM CHECKS

Service or change oil using the proper grade and type of oil. If oil brands are changed, follow the engine manufacturer's instructions for flushing the lubrication system.

Condition and operation of all chip detectors and chip detection warning circuits.

Oil filters for cleanliness and metal debris. If metal is detected, follow the engine manufacturer's instructions for methods to identify the type of metal to isolate possible sources.

Collect an oil specimen for a laboratory oil analysis as prescribed by the engine manufacturer.

Oil cooler for condition, cleanliness, and leakage. Check thermal by-pass valves for condition and operation.

Condition of oil cooler ram air ducts and the operation of control doors or shutters, when installed.

## Chapter 2

FUEL SYSTEMS

TURBINE ENGINE FUEL SYSTEMS

Introduction

The fuel system of the jet engine consists of

1. Fuel pump

2. FCU

3. Fuel heater

4. Fuel Deicing Unit

5. FCOC

6. P & D valve

7. Fuel nozzles

Some of these units are life based on flight hours/cycles. After completion of life, the units are to be overhauled and life starts from zero hours.

Before installing the fuel system components, this has to be ensured that they are serviceable. At the time of installation new packing are to be installed, all pipes lines to be crack tested and pressure tested.

This section describes the operation of typical turbine engine fuel systems and the maintenance normally carried out in service.

GENERAL

The Gael supplied to the combustion chambers of a turbine engine must be readily combustible and in a ratio with the mass air flow through the engine which will ensure efficient and economical operation under all conditions of flight.

Turbine Fuels

The fuels used in turbine engines must conform to rigid requirements to give optimum performance and safety. They must also be compatible with materials used in the fuel system components and provide adequate lubrication of working parts. The types most used in civil aircraft are to D. Eng. RD 2494 (A VTUR) or ASTM Spec. D1655 (Jet A or A-I) all of which are kerosene type fuels.

One problem associated with kerosene fuels is the water taken into solution through the aircraft or storage, tank venting system. This water may freeze at high altitude and result in the low pressure fuel filter becoming blocked with ice crystals. Fuel heaters are therefore necessary and are incorporated in most aircraft fuel systems upstream of the low pressure filter.

Fuel System Operation

The required engine speed is set by a throttle valve which passes a fixed amount of fuel to the spray nozzles (burners). Heat produced in the combustion chambers expands the gases rearwards to impinge on the turbine, resulting in rotation of the compressor/turbine assembly with the energy remaining in the gas stream providing. engine thrust. An increase in fuel 'flow results in higher temperatures and increased gas expansion, producing higher engine speed, greater airflow and increased thrust. The engine speed selected by the initial positioning of the throttle valve will be maintained provided that air intake conditions do not vary. Changes in altitude, air temperature and forward speed will affect mass air flow through the engine and a corresponding change in fuel flow is necessary to maintain the selected speed. In addition, any rapid throttle movements will uspet the air/fuel ratio due to the inertia of the compressor/turbine assembly. Automatic means are therefore necessary to relate fuel flow to mass air flow through the engine and to control maximum speed idling speed and acceleration rate. A convenient means of achieving these functions is to control output from the fuel pump.

FUEL PUMP

On some early turbine engines a constant displacement pump was used, the i,±sign of which ensured that pump delivery was always in excess of enginerequirements. Excess fuel was bled back to the fuel tanks by means of a unit called a Barostat which was, sensitive to changes in air intake pressure.

Variable delivery 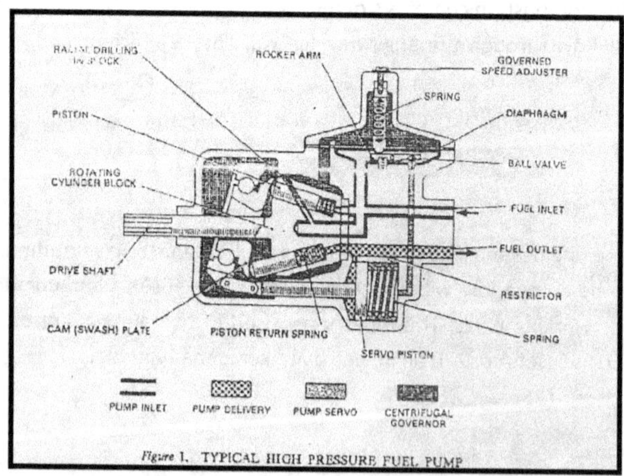

Figure 1. TYPICAL HIGH PRESSURE FUEL PUMP
pump

The variable stroke pump is driven directly from the engine and consists of a rotating cylinder block in which a number of cylinders are arranged around the rotational axis. A spring-loaded piston in each cylinder is held against a non-rotating cam plate so that rotation of the cylinder block results in the pistons moving up and down in their respective cylinders. Conveniently placed ports in the pump body allow fuel to be drawn into the cylinders and discharged to the engine. The angle of the cam plate determines the length of stroke of the pistons and, by connecting it to a servo mechanism, delivery may be varied from nil to maximum pump capacity for a given pump speed.

The servo piston operates in a cylinder and is subjected to pump delivery pressure on one side and the combined forces of reduced delivery (servo) pressure and a spring on the other. A calibrated restrictor supplies pump delivery fuel to the spring side of the piston and this is bled off by the control system to adjust the piston position and hence the angle of the cam plate.

FUEL PUMP CONTROL SYSTEMS

Some engines are fitted with a control system which uses electronic circuits to sense changing fuel requirements and adjust pump stroke. Most engines however, use hydro-mechanical systems, with an electro-mechanical element to control maximum gas temperature and these are discussed in the following paragraphs.

Pressure Control

The quantity of fuel passing through a restrictor (the throttle valve) may be varied by increasing or decreasing the fuel pressure. In the pressure control system fuel pressure is varied in relation to air intake pressure, decreasing with decreased mass air flow through the engine. Spill valves in the Barometric Pressure Control (B.P.C.), Acceleration Control Unit (A.C.U.) and pump governor, bleed off servo pressure to control pump stroke.

Under steady running conditions below maximum governed speed only the B.P.C. spill valve is open. A capsule subject to air intake pressure, contained in the B.P.C., controls the extent to which the spill valve is open. The bleed is arranged to increase as intake pressure decreases thus reducing servo pressure, pump stroke and fuel delivery pressure as altitude increases.

When the throttle is opened slowly, reduced throttle inlet pressure is transmitted to the B.P.C. and the spill valve closes to increase servo pressure and pump stroke. As pressure to the throttle is restored the B.P.C. spill valve again takes up its control-ling position, and pump stroke, combined with increased pump speed, stabilises to give the output for the new throttle position. If the aircraft is in level flight the increasing speed will increase intake pressure and act on the B.P.C. capsule to further increase fuel flow to match the increasing mass air flow.

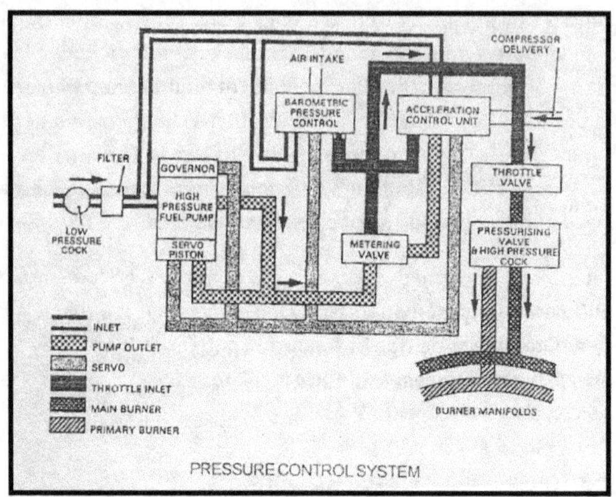

During rapid throttle opening, the action of the B.P.C. is restricted by the A.C.U. to prevent over fuelling as the B.P.C. spill valve closes increased fuel flow creates an increased pressure drop across the Metering Valve which is sensed by the A.C.U. fuel diaphragm. Movement of this diaphragm opens the A.C.U. spill valve to reduce servo pressure and limit overfuelling to the maximum amount which can be tolerated by the engine. As the engine accelerates, increasing compressor delivery pressure acting on the A.C.U. air diaphragm gradually closes the spill valve to permit greater acceleration at higher engine speeds.

Radial drillings in the fuel pump rotor direct fuel under centrifugal force to one side of a spring loaded diaphragm in the governor unit. When centrifugal force reaches a pre-determined value the diaphragm flexes sufficiently to open its spill valve and reduce servo pressure, thus limiting the amount of fuel delivered to the engine and so controlling engine speed.

Flow Control

In this system fuel pump delivery is controlled to maintain a constant pressure drop across the throttle valve regardless of engine speed. A common variation of the system is one in which a small controlling flow (proportional flow) is created with the same characteristics as the main flow and is used to adjust the main flow. A different type of spill valve known as a "kinetic" valve is used which consists of opposing jets of fuel at pump delivery pressure and servo pressure; a blade moving between the jets alters the effect of the high pressure on the low pressure. When the blade is clear of the jets, servo pressure is at maximum and moves the fuel pump to. maximum stroke but as the blade comes between the jets servo pressure reduces to shorten pump stroke. The control elements which are housed in a single unit called the Fuel Control Unit (EC. U.) are the Altitude Sensing Unit (A.S. U.), Acceleration

14

Control Unit (A.C.U.), Proportioning Valve Unit (P.V.U.) and throttle, which sometimes also functions as a shut-off (H.P.) cock. The system is illustrated in Figure.

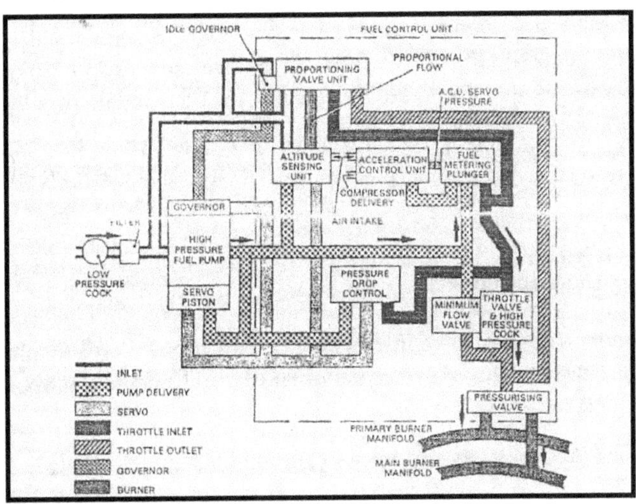

FLOW CONTROL SYSTEM

Under steady running conditions below governed speed, flow through two P.V.U. restrictors is proportional to flow through the throttle valve and the P.V.U. diaphragm is held open by spring pressure, allowing fuel to flow through the A.S.U. back to the pump inlet. The A.S.U. adjusts servo pressure in relation to this proportional flow by means of a kinetic spill valve.

When the throttle is opened slowly the pressure drop across the throttle valve and P.V.U. restrictors decreases and the P.V.U. diaphragm adjusts its position to reduce proportional flow through the A.S.U. This results in the A.S.U. spill valve closing slightly to increase servo pressure and therefore pump stroke, thus restoring the pressure difference across the throttle and P.V.U. restrictors.

Variations in air intake pressure arc sensed by a capsule in the A.S.U. which adjusts its spill valve to decrease or increase servo pressure as required. The resulting change in proportional flow returns the A.S.U. spill valve to its controlling position.

During rapid throttle opening the sudden decrease in pressure drop across thee is sensed .by the A.S.U. which closes its spill valve to increase pump stroke.

The rapid increase in fuel flow, which would cause over fuelling, is restricted by means of a pressure drop diaphragm and metering plunger. This diaphragm is sensitive to the pressure drop across the metering plunger, the latter being located in the main fuel line to the throttle valve. Rapid throttle opening increases the pressure drop across the plunger and at a fixed rate of over fuelling the pressure drop diaphragm flexes sufficiently to open its spill valve and override the A.S.U., maintaining a fixed pressure drop across the metering plunger. The metering plunger is, in effect, a variable area orifice and by means of a capsule in the A.C.U. sensitive to compressor delivery pressure, its position is controlled to increase the rate of over fuelling as engine speed increases. As the controlled over fuelling and engine speed increase, the pressure drop across the throttle valve is gradually restored until the proportional flow reaches a controlling value once more and the A.S.U. spill valve controls pump stroke.

Fuel under centrifugal force from the fuel pump also acts on a diaphragm in the P.V.U. to adjust the position of one of the restrictors and maintain proportional flow at a value suitable for idling.

Combined Acceleration and Speed Control

This fuel pump control system is contained within a single unit called a Fuel Flow Regulator, the fuel pump servo piston being operated by fuel pump delivery pressure opposed by main burner pressure and a spring. The system is illustrated in Figure.

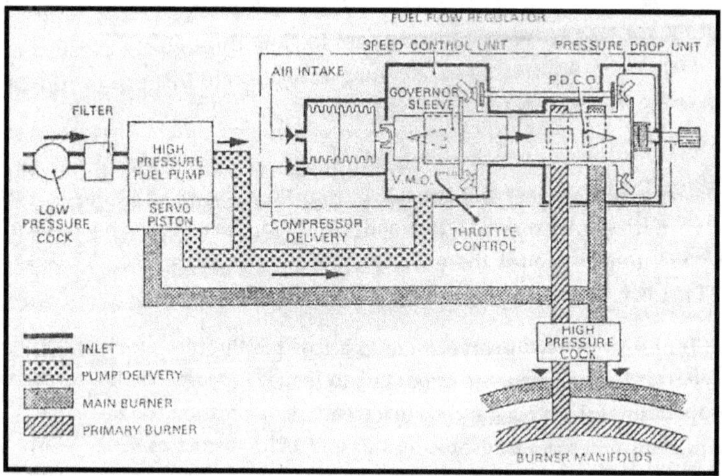

COMBINED ACCELERATION AND SPEED CONTROL SYSTEM

Two rotating assemblies, each with a hollow valve and centrifugal governor, are driven from the engine by a gear train in the regulator and are known as the Speed Control Unit and- the Pressure Drop Unit. The speed control valve is given axial movement by a capsule assembly under compressor delivery pressure and has a triangular hole known as the Variable Metering Orifice (V.M.O.). A non-rotating governor sleeve round this valve is given axial movement by the governor unit and restricts fuel flow through the V.V.O. Fuel from the pump outlet flows from the regulator body through the V.M.O. to the inside of the speed control valve and passes Through the hollow vale to pressure drop unit. Tile pressure drop valve is in the form of a hollow piston, moving axially under the force of fuel from the V.M.O. and governor flyweights, opposed by main fuel pressure. The pressure drop valve has an unrestricted outlet through the regulator body for primary burner fuel and a tri-angular outlet known as the Pressure Drop Control Orifice (P.D.C.O.) through which fuel flow to the main burners is restricted by the axial movement of the pressure drop valve.

Under steady running conditions the position of the speed control valve is fixed by the capsule assembly and the governor sleeve is held in a fixed axial position by the speed control governor. Pressure drop across the V.M.O. is sensed by the pressure drop valve which adjusts its position and the exposed 'area of the P.D.C.O. to supply the correct quantity of fuel in relation to engine speed.

When the throttle is opened slowly, spring loading on the speed control governor increases to move the governor sleeve and increase the V.M.O. area. Pressure drop across the V.M.O. decreases and this is sensed by the pressure drop valve which moves to increase the size of the P.D.C.O., the reduced system pressure difference acting on the pump servo piston to increase fuel flow to the engine. As the engine accelerates, the capsule in the speed control unit is compressed by increasing compressor delivery pressure and moves the speed control valve to further increase the size of the V.M.O. Balance is restored when centrifugal force acting on the speed control governor moves the governor sleeve to restore the system pressure difference.

The effect of rapid throttle movements is restricted by mechanical stops acting on the governor sleeve.

Changes in altitude or forward speed affect the capsule in the speed control unit which adjusts the position of the speed control valve to correct fuel flow.

BURNERS

The purpose of the burners is to provide fuel to the engine in a suitable form for combustion. A burner with a single spray nozzle, although used on some early engines, is not suitable for large modem engines due to the widely varying fuel flow requirements for different flight conditions. If the orifice were of a size suitable for atomising fuel at low rates of flow the pressure required at take-off would be tremendously high, flow through an orifice being proportional to the square of the pressure drop across it.

One of the methods used to overcome this problem is the provision of a dual spray burner. The central orifice provides the fuel for low how rates and a second annular orifice is used in addition for high flow rates. Distribution between the primary (low flow) manifold and the main (high flow) manifold is normally controlled by a pressure operated valve. In the case of the Fuel Flow Regulator, fuel flowing through the rectangular outlet from the pressure drop valve is always at a higher pressure than fuel flowing through the outlet from the P.D.C.O and is used to supply the primary burner manifold.

Another method used on some engines is known as the Vaporising Burner. Fuel is injected at low pressure into one end of a hollow "U" shaped tube located in the combustion chamber. It mixes with the primary air flow, is vaporised by the heat in the chamber and ejected upstream into the combustion zone. In this system a separate burner is necessary for engine starting.

ADDITIONAL CONTROLS

In systems additional controls are usually provided to prevent the engine from exceeding operating limitations.

Turbine Gas Temperature Control

Control of the maximum permitted turbine gas temperature is often exercised electrically. Signals from the T.G.T. thermocouples are amplified to either actuate a solenoid operated valve in the fuel system or reset the throttle linkage to reduce fuel flow to the burners. On engines which have different T.G.T. limitations for climb and take-off, a switch on the flight deck presets the T.G.T. signal reference datum.

Compressor Control

In certain circumstances such as high forward speed and low ambient temperature it is possible to produce maximum power/ thrust at less than maximum engine speed.

Under these conditions the engine could sustain damage due to high compressor delivery pressures and fuel flow is restricted by providing a bleed from the A.C.U. capsule chamber to atmosphere when compressor delivery pressure exceeds a precieLerinine.ci value.

To prevent the low pressure compressor from exceeding its design speed a centrifugal governor driven from the low pressure shaft is often included in the fuel system. If design speed is exceeded the low pressure governor restricts the fuel flow in the main burner line and reduces both high and low pressure compressor speeds.

Maintenance and Inspection

Fuel systems normally operate at very high pressures and a leak from a connection could quickly become a potential fire hazard. Regular inspection of the complete system for security, bonding, freedom from leaks and correct positioning of drains is therefore very important. If a leak does develop the tightness of the connection should be checked but on no account should the recommended torque values be exceeded. If the leak persists the connection should be dismantled, the joint faces inspected for cleanliness, and the connection remade using new gaskets, washers or '0' rings as appropriate. Pipes which are found to be twisted or damaged, particularly at the connections, must be replaced.

Lubrication of working parts should be carried out in accordance with the approved Maintenance Schedule using the type of lubricant recommended by the manufacturer. All excess oil or grease should be removed. Lubrication of gaskets, washers and '0' rings is usually required during assembly of components. It is important that only the correct type of lubricant is used, as incorrect lubricatio1 could cause rapid deterioration-of the seals, resulting in fuel leakage with the attendant fire risk. Whenever a component is replaced within the fuel system it is good practice to also clean the main fuel filter, especially where it incorporates a relief valve designed to open in the event of the filter becoming blocked. The surge of fuel on switching on the booster pump with a dry component in the line can open the relief valve and pass dirt into the system.

Cleanliness

Cleanliness of the fuel and fuel components is very necessary because of the small passages, restrictors and valves used in the system. Particular care should be taken to prevent the introduction of extraneous material such as water, dust or grit into the fuel tanks when refuelling an aircraft. Refuelling nozzles should be kept scrupulously clean. Filters which are incorporated in the supply line to the fuel pump and other small filters which may be fitted at the inlet connection of other components should be removed for examination at the intervals prescribed in the approved Maintenance Schedule. Except where renewable elements are used these filters should be thoroughly washed in kerosene and dried with compressed air before refitting. The low pressure fuel filter is normally provided with a cll'ain valve for fuel sampling purposes.

When changing components all pipes and open passages should be blanked off immediately and the blanks removed only when the new component is ready to be fitted. New washers, gaskets or '0' rings should always be refitted with new components.

General

Adjustments to the fuel system may be necessary when a new engine is installed, when a component is replaced or when incorrect operation of the system in flight is reported. Each component is bench tested before final approval and it is usual for the manufacturer to limit the extent to which adjustments can be made in service. If it becomes apparent that the permitted adjustment would be exceeded in order to achieve correct engine operation then the appropriate air and fuel pipes must be checked for leaks. If the required adjustment is still excessive it must be assumed that the component concerned is unserviceable and it should be replaced. A record of any adjustments made should be entered in the engine log book to provide a history of fuel system operation.

As air temperature, pressure and humidity all affect engine operation, any adjustments to the system must take account of ambient conditions and of the specific gravity of the fuel. Graphs or tables are provided in the appropriate Maintenance

Manual showing the correction to be applied to a basic setting under different operating conditions. For example, when setting an idling speed, which may be 40 % at standard atmospheric pressure of 1013 millibars, it may be necessary to reset the actual engine speed to 43.5 % when the atmospheric pressure is only 950 millibars.

Mechanical Adjustments

Adjustments to the idling speed, maximum governed speed and acceleration rate are usually by means of a screw and the effect of a set amount of rotation of the screw is usually quoted in the Maintenance Manual to avoid unnecessary engine running. Before making an adjustment the engine should be run for a sufficient length of time to stabilise engine temperatures and when altering power settings the controls should be operated carefully, and instruments kept under observation, to prevent exceeding operating limitations. This is particularly important after the installation of a new engine or component change. Adjusters must be relocked after use.

Electronic Adjustments

Some automatic engine controls are actuated by signals from the jet pipe thermocouples and require special test sets to check or correct their operation. Control is effected by comparing the thermocouple voltage with a datum voltage preset in an amplifier. Excessive temperatures produce an excess voltage and this is amplified to energise a relay in the T.T.C. circuit. This action operates a solenoid located either in the Fuel Bleed Valve or Flow Control Unit and restricts fuel flow to the engine. To adjust a control system of this type it is necessary to attach a special test set to the electrical circuit by means of a conveniently placed multi-pin plug and socket. The test set feeds a voltage to the circuit equivalent to that which would be supplied by the thermocouples at a specified gas temperature and any variations from the datum voltage may be reset by adjustment of resistors in the engine amplifier. Engine installations vary considerably in the layout of the automatic control systems and the appropriate Maintenance Manual should be referred to whenever it becomes necessary to check or adjust a specific system.

Control Linkage

The methods used to connect the pilots' controls to the engine throttle valve and the high and low pressure cocks an often very complex. The linkage normally includes '41*- cable and chain components in the flight deck and fuselage and a series of push/ pull rods and levers leading to the engine bay. There are numerous connections in the installation and this can lead to 'lost motion' at the pilots controls and irregular engine operation when wear takes place. The procedure for initially setting up the controls i3 laid down in the appropriate Maintenance Manual and the operation requires the use of rigging pins, datum marks, protractors and pointers. Readjustment may be required when wear in the connections causes slackness and insufficient travel of the control. At the periods specified in the Maintenance Schedule the linkage should be lubricated and. inspected for security, locking, play and correct adjustment. Excessive wear should be eliminated by replacement of the affected parts.

Pipes

When changing fuel system components or pipes care is necessary to ensure that the pipes are not strained as this could result in the development of leaks due to the high fuel pressures in the system: Short rigid pipes are often used between components and special fitting instructions may be quoted in the Maintenance Manual. It may, for instance, be necessary to loosen or remove an adjacent component in order to fit a pipe. When installing a flexible pipe care must be taken to prevent the pipe from twisting when the connections are tightened. Before pipes are reconnected they should always be inspected for cleanliness and damage, especially at flared ends and nipples. Seals should be lubricated in accordance with the manufacturers recommendations and union nuts must be correctly locked. Whenever pipes have been disconnected the system should be bled.

Fuel Pump

It is possible to cause damage to the fuel pump when shutting clown an engine in flight. Provided that the H.P. cock is closed first no damage will be caused by a wind milling engine but if the engine is stopped by closing the L.P. cock, or if the pump inlet is blocked in any way, the engine manufacturer usually specifies a time limit on engine wind-milling as the pump will be running "dry" and damage may result. If this time Limit is exceeded the pump must be changed and an inspection for contamination of the system downstream of the pump carried out. If metal deposits are found all system components must be changed and connecting pipes flushed out.

Burners

Due to their position in the combustion chambers, burners may contaminated by deposits of carbon which could affect their operation. The deposit r should be removed at the intervals laid down in the Maintenance Schedule by either one of two methods depending on whether or not the burners can be removed from the installed engine. On no account should a wire brush be used to remove carbon, as any scratches will affect the spray pattern and result in hot spots in the combustion chamber.

In-situ cleaning, where approved, is carried out by connecting a pumping rig to the burner manifold feed pipe connection and pumping a set quantity of carbon solvent through the burners. After removing the pumping rig and reconnecting the feed pipe, the engine should be left for approximately two hours to give the solvent time to soften the carbon and then run for a short time to disperse the deposit.

When the burners are easily removed from the engine they should be completely immersed in a carbon solvent for two hours then thoroughly cleaned with an air/ water gun, dried and dipped in de-watering oil. They may then be replaced in the Ir. engine or, if not required for immediate use, filled with inhibiting oil, blanked off and stored.

NOTE: Burners are often kept in sets and should only be replaced as individual components when permitted by the manufacturer.

Bleeding

Whenever a component is replaced or disconnected and instability in engine speed which could be attributed to air in the system is encountered, the fuel system should be bled. Fuel supply to the affected engine should be selected on and the low pressure cock and fuel pump turned on. Individual components should then be bled in turn commencing at the H.P. fuel pump and working downstream, each bleed valve being opened until bubble-free fuel is discharged, then the valve closed and locked. On some components a special bleed tool is provided and the discharged fuel should be caught in a container to avoid contamination of the engine bay.

Inhibiting

Except for separate electrical parts all components removed from the fuel system should be inhibited as soon as possible to prevent internal corrosion. All fuel should be drained out and inhibiting oil poured in through the inlet connection until full. A blank should then be fitted and the component rotated, topping up through each outlet in turn and fitting blanks until it is completely full of oil and securely sealed. Drive shafts of pumps and 'fuel flow regulators should be turned while inhibiting and the throttle valve and H.P. cock operated several times to ensure complete distribution of oil. To check that the blanks are not leaking the part should be thoroughly dried and left for thirty minutes. Any leaking blanks should be replaced, the oil level topped up and a further test carried out until satisfactory results are obtained. Drive shafts should then be smeared with the recommended storage oil.

NOTE: The inhibiting oil-can should contain a fine mesh filter to prevent the ingress of foreign material.

Packing

All components, including the electrical ones which are not inhibited, should be wrapped in greaseproof paper, sharp edges being double wrapped, then enclosed in V.P.I. paper and secured. with adhesive tape. The wrapped parts, together with a label giving details of the modification state, reason for return, etc., should be enclosed in a polythene bag, as much air as possible excluded and the bag heat sealed. Transport boxes are normally provided for all components and the wrapped part should be kept in one of these during storage. A label should be affixed to the box giving details of the contents, inhibiting date and storage life.

NOTE: Certain electrical components require the attachment of a "Fragile" label on the transport box in addition to the normal identification label.

Storage Conditions

Components should be stored in conditions that are dry and free from corrosive fumes. Components are best stored in racks but this may not always be possible; a double fuel pump, for example, may weigh up to 75 lb without its transport box and may more conveniently be stored at floor level but raised on blocks to permit the circulation of air.

Every six months the shafts of fuel pumps and fuel flow regulators should be rotated a few turns without removing them from the sealed bag. At the same time the part may be checked to ensure that the modification state is satisfactory and that no leaks have developed.

Some fuel system components may be given a maximum storage life. When this life has expired the components should be removed from storage and subjected to such overhaul and testing as may be specified by the manufacturer.

TURBINE ENGINE FUEL CONTROL MAINTENANCE

Turbine engine fuel control repairs in the field consist of control replacement or occasional field adjustments. Furthermore, adjustments are limited to idle and maximum speed adjustments. If you recall, the process of adjusting a turbine engine fuel control is commonly referred to as trimming the engine. The primary purpose for trimming a fuel control unit is to ensure the availability of maximum thrust output when needed.

Trim checks are completed whenever engine thrust is suspect, and after such maintenance tasks as prescribed by the manufacturer. An engine change, fuel control change, or throttle linkage adjustments for proper control cushion and spring back are all examples which require trimming procedures. A fuel control may also need to be re-trimmed when deterioration of engine efficiencies occur as service time takes its toll. Another example is when wear and tear on engine control linkages cause misalignment between the cockpit and engine.

Manual trimming procedures vary widely between engine models; therefore, before you attempt to trim an engine, you should take time to review the specific procedures in the engine's maintenance manual. A typical trimming procedure requires you to install calibrated instruments for reading turbine discharge pressure or EPR. In addition, a calibrated tachometer must be installed to read N2 rpm. Once the instrumentation is installed, the aircraft should be pointed into the wind. However, if the velocity of the wind blowing into the intake is too great, elevated compression and turbine discharge pressures will resist which, ultimately, will produce a low trim setting. Another step in the trimming procedure is to measure the barometric pressure at the engine inlet and the ambient temperature. This is required to correct performance readings to standard sea-level conditions. To obtain a temperature reading it is common practice to hang a thermometer in the shade of the nose wheel well. The ideal conditions for trimming a turbine engine are no wind, low humidity, and standard temperature and pressure.

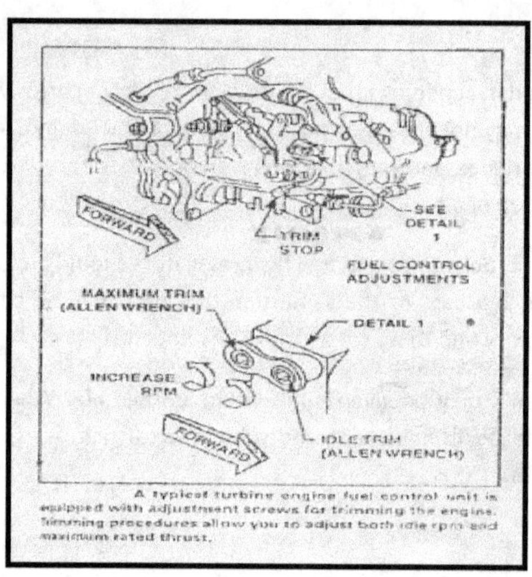

A typical turbine engine fuel control unit is equipped with adjustment screws for trimming the engine. Trimming procedures allow you to adjust both idle rpm and maximum rated thrust.

Once you have calculated the maximum power output for the engine you are trimming, start the engine and let it idle as specified in the maintenance manual. This is the engine has time to stabilize. To ensure an accurate trim setting, all engine bleed air must be turned off. Bleeding air off the compressor has the same effect as decreasing the compressor's efficiency. Therefore, if a trim adjustment is made with the bleeds on, an inaccurate, or over trimmed condition will result.

With the engine running at idle and maximum power, observe the turbine discharge or EPR reading to determine how much trimming is necessary If trimming of either the idle or maximum settings is necessary, it is typically accomplished by turning a screw type adjustment on the fuel control unit. However, most manufacturers recommend that in order to stabilize cams, springs, and linkages within the fuel control, all final adjustments must be made in the increase direction. If an over adjustment is made, the trim should be decreased below target values, then increased back to the desired values.

Typical trim adjustments only require simple, hand tools such as an Allen wrench, screwdriver, or wrench. Adjustment components are usually equipped with some sort of friction lock that does not require the use of lockplates, locknuts or lock-wire. Some engines are designed to accommodate remote adjusting equipment. A remote control unit allows you to make the trim adjustments from the cockpit during ground test with the cowls closed.

Another important part of the trim procedure is to check for power lever cushion, or spring-back. You should move the power lever full forward and release it before and after the trim run. The amount of lever spring-back is then measured against prescribed tolerances. If the cushion and spring-back are out of limits, you must make the necessary adjustments in accordance with the manufacturer's rigging instructions. Correct power lever spring-back ensures a pilot that takeoff power will be obtained and that additional power lever travel is available for emergencies. Correct spring-back is indicated when the fuel. control reaches its internal stop before the cockpit power lever reaches its stop.

A trim check is normally followed by an acceleration check. After completing the trim check, place a mark on the cockpit power lever quadrant at the take-off f trim position. Then, advance the power lever from the idle position to take-off thrust position and measure the time against a published tolerance. A typical acceleration time from idle to take-off thrust for a large gas turbine engine ranges from 5 to 10 seconds.

Once the fuel control unit has been trimmed and control spring-back is correct, the engine should produce its rated thrust in standard conditions. In non-standard conditions with high ambient temperatures, the rated thrust is degraded. While low humidity is desirable for purposes of accuracy during trimming procedures, high atmospheric humidity actually degrades rated thrust very little. The reason for this is that only 25 percent of the air passing through a turbine engine is used for combustion.

## Chapter - 3

IGNITION SYSTEM

Starting and Ignition System

Two separate systems are required to start a turbine engine, a means to rotate the compressor/ turbine assembling and method of igniting fuel - air mixture in the combustion chamber. The starter motor is capable of cranking the engine to a speed higher than at which sufficient gas flow in generator to enable the engine to accelerate under its power. During cranking operation, ignitor plugs provided in the combustion chamber supplied with electrical power, followed by the injection of fuel when fuel pressure has built up sufficiently to produce and atomised spray. Light up normally occurs at this point and the engine assisted by the starter motor, accelerates to self-sustaining speed. The starter drive disengages when engine power begins to sustaining speed. Power supplies to-the starter and ignitors are cancelled during this start. During starting of the engine, it is most important not to exceed the exhaust gas temperature by cutting off the fuel supply to the engine. There are different methods used to crank the engine to self-sustaining speed. An air starter is most commonly used on passenger transport aircraft as this is an economical means of starting, causing the minimum disturbance. to the passenger. Electrical starters are also commonly used and are fitted mainly turbo prop and small jet engines.

Certain operators may used to start aircraft engines without outside assistance and specify the use of a hydraulic or air starter driven from an APU.

On some early engines, a torch ignitor which combined a sparking plug and fuel spray nozzle was used during starting, fuel being supplied by a priming pump controlled by a timer unit.

To provide automatic relighting in the event of flame extinction either a glow plug or continuously operated ignition system are used as manual relighting is accomplished by operating the ignition independently from the complete starter circuit. On some aircraft the ignition circuit is also corrected to a stall warming device as a safe guard against flame extinction under stall conditions.

During installation of oil system components to the engine, required installation procedures to be followed as per overhaul manual.

The packings or O rings to be ensured for having a shelf life, and lubricated properly with approved lubricant during installations. After installations of components, check for oil leaks after brief ground run.

While rectification of oil system snags it is advisable to have oil flow diagram to trace the oil system problem.

Component Replacement

The following inspection requirements should be satisfied whenever it becomes necessary to change an unserviceable item.

(i)   The new component should be inspected for damage or corrosion, particularly on mating surfaces, and if possible its operation checked before installation.

(ii)   On components with electrical mechanisms the insulation resistance between each pin of the electrical sockets and the case should be tested with a 250 volt insulation resistance tester, the minimum value required being 50000 ohms. The aircraft electrical supply should be disconnected by removing the appropriate fuse before installing a component of this type.

(iii)   When the incorrect installation of a valve could cause malfunction or damage to the system, the component is usually designed so that it can only be fitted in one way. When this is not practical an arrow may be embossed on the casing to show the direction of main flow and it is important to take note of any such marking and refer to the Maintenance Manual when the installation procedure is not obvious.

(iv)   "V" flange clamps are often used to connect components together and lubrication of the mating faces is sometimes required by the manufacturer. On other types of connection the use of jointing compound may be specified. All bolts should be torque loaded to the value specified in the Maintenance Manual.

(v)   New washers, gaskets or seals should always be used when replacing a component in the system.

Testing after Component Replacement

Whenever a new component has been installed a functional check should be carried out to ensure correct operation of the system.

(i)   Operation of the starter motor may be checked by carrying out a motoring run.

        During this test the operation of each valve should be checked against its indicated position and the engine speed obtained should be within the limits quoted in the Maintenance Manual.

(ii)   If a valve concerned in the operation of both the starting and pneumatic systems is changed (e.g. an isolation valve), it will be necessary to start the engine and check the operation of both systems. Satisfactory cranking speed and pneumatic duct pressures should be obtained.

IGNITION SYSTEMS

The ignition system of a turbine engine must provide the electrical discharge necessary to ignite the air/fuel mixture in the combustion chamber during starting and must also be capable of operating independently from the starter system in the event of flame extinction through adverse flight conditions. The electrical energy required to ensure ignition of the mixture varies with atmospheric and flight conditions, more power being required as altitude increases. Two independent 12 joule systems are normally fitted to each engine to provide a positive light up during starting but some engines have one 12 joule and one 3 joule system. The 3 joule system is kept in continuous operation to provide automatic relighting.

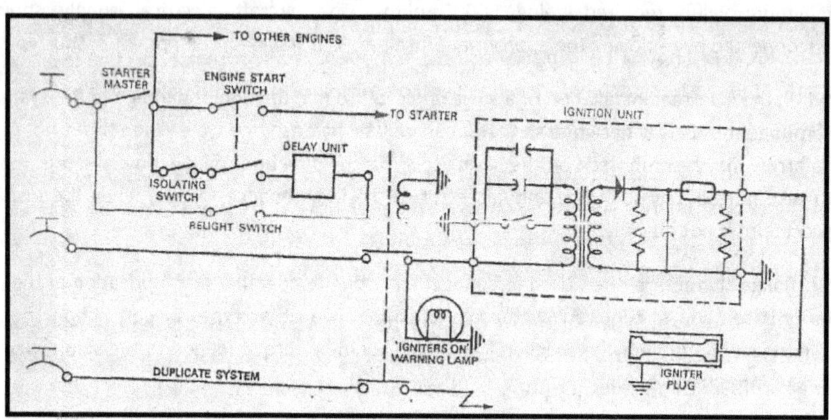

TYPICAL. IGNITION SYSTEM

Where continuous operation of one system is not desirable, a glow plug is sometimes fitted in the combustion chamber where it is heated by the combustion process and remains incandescent for a sufficient period of time to ensure automatic re-ignition.

High Energy Ignition Unit

A 12 joule unit receives electrical power from the aircraft d.c. supply, either in conjunction with starter operation or independently through the "relight" circuit. An induction coil or transistorised h.t. generator repeatedly charges. a capacitor in the unit until the capacitor voltage is sufficient to break down a sealed discharge gap. The discharge is conducted through a choke and h.t. lead to the igniter plug where the energy is released in a flashover on the semi-conducting face of the plug. The capacitor is then recharged and the cycle repeated approximately twice every second. A resistor connected from the output to earth ensures that the energy stored in the capacitor is discharged when the d.c. supply is disconnected.

NOTE: A 3 joule unit is usually supplied with h.t. alternating current but its function is similar to that described above.

The electrical energy stored in the high energy ignition unit is potentially lethal and even though the capacitor is discharged when the D.C. supply is disconnected, certain precautions are necessary before handling the components. The associated circuit breaker should be tripped, or fuse removed as appropriate, and at least one minute allowed to elapse before touching the ignition unit, high tension lead or igniter plug.

Ignition units are attached to the aircraft structure by anti-vibration mountings and the rubber bushes should be checked for perishing at frequent intervals. It is also important that the bonding cable is securely attached, making good electrical contact and of sufficient length to allow for movement of the unit on its mountings.

Igniter Plug

The igniter plug consists of a central electrode and outer body, the space between them being filled with an insulating material and terminating at the firing end in a semi-conducting pellet. A spring-loaded contact button is fitted at the outer end of the electrode. During operation a small electrical leakage from the ignition unit is fed through the electrode to the plug body and produces an ionised path across the surface of the pellet. The high intensity discharge takes place across this low-resistance path.

Igniter plugs should be inspected at frequent intervals for security, damage, gas leakage and secure attachment of the h.t. lead. When removed they should be inspected for heat damage, cracks and erosion of the pellet surface. Igniter plugs are not normal cleaned but if carbon deposits make inspection of the pellet impossible the carbon may be removed, care being taken not to damage the surface of the pellet.

When it is necessary to fit a new igniter plug the manufacturer sometimes specifies that the depth of penetration of the plug into the combustion chamber should be checked. This is accomplished by means of a special tool similar to a dummy plug and the adjustment is made by selecting a shim of appropriate thickness to fit tinder the igniter plug housing. A new sealing washer must be fitted when a plug is replaced.

Lubrication of plug threads is normally specified by the manufacturer and plugs should be torque loaded to the value stated in the appropriate Maintenance Manual.

Ignition Lead

The high energy ignition lead is used to carry the intermittent high voltage outputs from the ignition unit to the associated igniter plug. A single insulated core is encased in a flexible metal sheath and terminates in a spring-loaded contact button at each end. The end fittings usually incorporate a self-locking attachment nut.

Before installing an ignition lead the spring-loaded contact assemblies should be checked for freedom of movement and, where specified, an insulation resistance check carried out in accordance with the appropriate Maintenance Manual. The sheath should also be checked for fraying and the ceramic insulating sleeves for cracks or other damage. The manufacturer may specify the use of an anti-seize' compound on the plug threads during fitting.

During service the leads should be inspected for security and damage. In particular the sheath should be examined in the vicinity of supporting clips for signs of chafing and over its whole length for signs of oil contamination.

Testing

Whenever an ignition component is changed or incorrect operation of the system is suspected a functional check may be made by operating the relight circuit.

The aircraft should be located in the open air and the engine inspected for signs of fuel or fuel vapour which, if present, must be dispersed before operating the ignition units. A suitable CO_2 fire extinguisher should be positioned adjacent to the engine before carrying out the test.

The high pressure fuel cock should be closed and the circuit breaker tripped (or fuse removed if appropriate) from each of the ignition circuits in turn whilst checking operation of the other. When the necessary switches are set for relighting, operation of the ignition system will be heard as regular clicking noises from the igniter plug as the electrical discharges occur and, on some aircraft, shown by illumination of an "igniters on" warning lamp on the flight deck. If a component common to both the ignition and cranking systems is changed (e.g. a time delay unit), it is advisable to carry out an engine motoring run with the H.P. cock turned oft- to check the normal D.C. circuit to the ignition units.

STORAGE

Starters and ignition components should be stored in conditions that are clean, dry, warm and free from corrosive fumes. A temperature of 16°C (61°F) and humidity of 75% are often quoted by manufacturers as being ideal storage conditions.

Starters and Ignition Units

These components are transported in either a "tropical" or "commercial" pack. "Tropical" packing includes sealing the component with a suitable quantity of desiccant in a polythene. bag and placing the bag in a padded wooden or cardboard box. Grease-resistant paper is used instead of a polythene bag for "commercial" packing. Components should be kept in their, boxes during storage and when unpacked for use the packaging material should be retained. If the original material is not available a returned component should be packed in accordance with BS 1133 or equivalent specification

Starter gearboxes should be drained before packing and external threads and drive shafts coated with a rust preventative.

Igniter Plugs

Plug threads should be coated with a rust preventative and the plug wrapped in waxed paper or a polythene tube. It may then be placed in a cardboard box, either singly or with other plugs, and the box sealed.

Ignition Leads

These should be wiped with a cloth moistened with white spirit to remove any oil or grease, and the end connections blanked off. The leads should be placed in a natural position not coiled or bent) on a flat shelf and covered with a dust cloth.

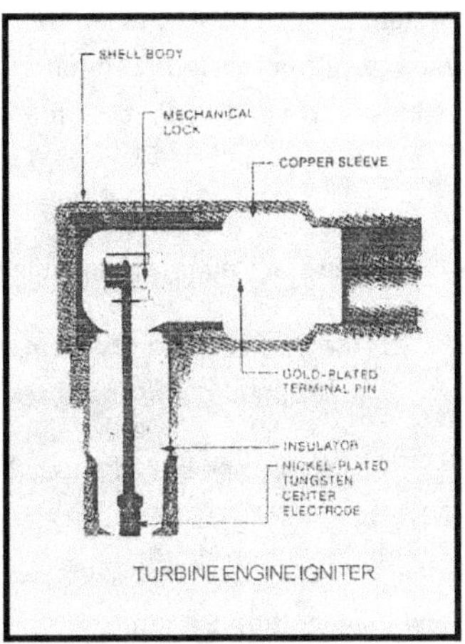

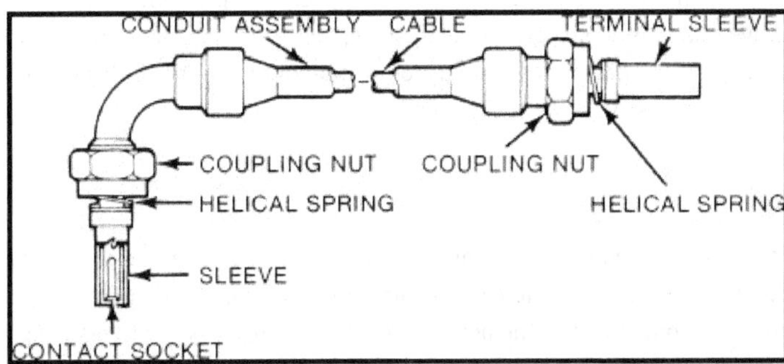

Storage Life

A specific storage life may be recommended for some components but igniter Plugs and leads may normally be kept in storage indefinitely provided that storage conditions are ideal. Any component which has reached the end of its storage life must be subjected to such inspection and testing as may be specified to enable re-certification for a further period of storage.

Auto-Ignition

Many turbine powered aircraft have an auto-ignition circuit installed. It is designed primarily to ensure instantaneous ignition if an engine begins to lose power from inlet icing and is in danger of flaming out. When the auto-ignition arming switch is "On" in the cockpit, electrical power is supplied to a sensor on the engine which is alerted if the engine experiences a sudden reduction in power. The two most common sensing systems utilize either loss of torque oil pressure or loss of compressor discharge pressure to actuate the auto-ignition system and to illuminate a cockpit warning light.

The next generation Boeing 737, like the 737 - 800 has an automatic ignition feature. On this airplane, auto ignition is enabled when the engine starts switch is in the "off" position, versus having a separate switch position labeled automatic. The ignition system will be automatically activated by the electronic engine control (EEC) when the following conditions are used.

1.	The start lever is in the idle position:

2.	The EEC sees an engine deceleration (without command) greater than normal schedule or the core speed is below 57% N2;

3.	The core speed is greater than 40% N2

Glow Plug Ignition System

A glow plug system, used on some models of gas turbine engines, provides current to a hot coil element in each glow plug which makes the plug glow red hot slow plug reaches an extremely high temperature very rapidly and ignites the fuel spray in the combustion section of the engine during starting.

Ignition System Checks

Condition and operation of ignition system elements including cleanliness and wear limits of igitilers or glow plugs as appropriate. NOTE. Always follow manufacturer's warnings when checking the operation of any ignition system component due to the potential for serious injury or death resulting from electric shock

a.	Condition, routing and security of ignition system leads and shielding.

b.	Security of attachment and condition of ignition regulators and igniter boxes.

c.	Operation of continuous-duty and auto-ignition modes.

Electrical and Starting System Checks

a.	Check condition and security of all electrical wiring and connectors inside the engine nacelle.

b.	 Inspect condition and security of pneumatic starters -or electric starter/ generators.

c.	Inspect for electric starter/generators check wear limits and condition of brushes.

d.   Inspect wear and condition of starter-generator spline drives and other drive elements such as shear shafts and couplings.

e.   Check condition and security of starter relays and solenoids.

f.   Inspect output of electric power generating components during pre- and post-inspection run-up checks. For multi-engine aircraft, check parallel operations from all electric power generating systems for equal load distribution and fault isolation capability in the event of a failed unit.

g.   Inspect for pneumatic starting systems, check operation of bleed air cut-off and start control valves.

TROUBLESHOOTING IGNITION SYSTEM

PROBLEMS / POSSIBLE CASE	CHECK PROCEDURE	REMEDY
1. No igniter spark with the system turned on		
a. Ignition relay	Correct power input to transform unit	Correct relay problems, refer to starter generator circuit
b. Exciter Unit	Correct power output	Test with special equipment observing ignition system cautions
c. High tension lead	Continuity or high- resistance shorts with ohmmeter and meggar test unit	Replace lead
d. Igniter plug	1. Damaged insulator or damaged semi conductor	Replace plug
2. Long interval. between sparks.		
a. Power supply	Weak battery	Recharge battery
3. Weak (low intensity) spark		
a. Igniter plug	Cracked ceramic insulation	Replace plug

## Chapter - 4

STARTING SYSTEM

Starting and Shutdown Faults

Aside from internal engine problems, inappropriate starting techniques and other problems that are external from the engine may cause starting malfunctions. When attempting to start a turbine engine for maintenance checks, it is advisable to always use auxiliary pneumatic or electric power sources. Insufficient starter power may prevent the engine from achieving proper compressor rotational speeds to support normal combustion. In addition, whenever starting a turbine engine, you should also consider ambient temperature and wind conditions. If the engine and fuel has been cold soaked from exposure to extremely cold outside temperatures, it may not be possible for the fuel to properly atomize to support combustion. In extremely cold climates, consider moving the aircraft into a hangar for preheating,

Strong winds blowing through the exhaust duct of an engine may also impede normal air or gas flow through the engine. If the engine fails to start in high wind conditions, reposition the aircraft and perform a dry motoring run to purge residual fuel from the combustors before attempting another start.

Once you have determined that there is sufficient power for starting and after eliminating all other external factors, begin systematically isolating possible engine problems that may be affecting the start. Typical starting problems caused by engine and engine accessory faults consist of the following:

1.   Does the starter motor operate? Audible sound confirm that can hear the starter during the start cycle. If there is no indication of starter operation, check the following:

Check electrical or pneumatic sources to determine that they are adequate and delivered to the starter. When cross-bleeding pneumatic air from a secondary source such as an APU or another engine, check the operation and position of all bleed-air and start control valves. For electric starters, check the operation of all starter relays and contactors, and the condition of electrical wires and terminal ends.

2.   If the starter operates, does the engine rotate? This may be confirmed with the compressor tachometer (Low pressure compressor or N1, high pressure compressor or N2, or gas generator compressor or Ng, as appropriate to the type of engine) or by visually checking for engine rotation.

3.   If there is no rotation, or if rotation is sluggish; check the starter driveshaft for condition and engagement. Manually attempt to rotate the engine by hand and listen for abnormal scrapping or rubbing sounds. Determine the source of any irregularities.

4.   If the engine rotates to an appropriate rpm, but still fails to start, check the following items:

Inspect the ignition system for proper operation. Igniters produce a loud snapping sound during starting. Glow plug ignition systems can be checked by removing the plugs from the combustor and turning the ignition on to check heating elements. If ignition is not evident, check power supplies or the ignition exciter. Also perform insulation and continuity checks on ignition cables.

Check the fuel pumps, FCU, and fuel nozzles for proper operation and condition. inspect all aircraft and engine fuel filters and screens for contamination. Check the fuel control unit and fuel pumps for proper output pressures, and inspect nozzles for cleanliness and flow check as necessary.

Using a fiberoptic borescope, perform an engine inspection to check for severely damaged internal hot section and compressor components.

5.   If the engine starts, but tends to hang up, or is slug-gish to accelerate, check the following items:

check all pneumatic bleed air reference lines to the fuel control unit (FCU) and other pars of the engine for leakage, or blockage.

Check starting bleed air valves for proper operation. Often, bleed air valves remain open during the initial part of the start sequence to reduce the load on the compressor, but close at higher rpm before light off. Also ensure that all auxiliary airframe system bleed air valves are properly positioned during the start.

Check the pressurization and dump valve (P&D or purge valve) for proper operation. After shutdown, the valve should open to allow excess fuel to drain, but should close during the start sequence.

Perform additional checks as required for the specific engine as called for in the manufacturer's troubleshooting procedures.

Operational Faults

Operational faults include items that become apparent during normal operations. Faults may include failure of the engine to accelerate or decelerate properly, compressor surges or stalls, transient over temperature conditions, abnormal vibrations, and others.

Engine indicating systems sometimes erroneously show values that are outside normal operating parameters. You should always verify that the engine indicating systems are calibrated and working properly before beginning to diagnose internal engine faults. A Jetcal engine analyzer or similar equipment, should be used to check the integrity of the indicating gauges, transmitters, sensing elements, and wiring harnesses.

Some typical operational faults and actions to take to diagnose them are as follows:

1.   During mid to low power operations, if a hooting or huffing sound is emitted:

Check for excess compressor air leakage through bleed air valves. In some instances, manufacturers may recommend blanking off bleed air outlets to diagnose whether the problem is caused by an engine bleed air problem, or possibly that the problem is leakage in an airframe bleed air (pneumatic) system.

Check the condition of the compressor inlet, blades, and guide vanes for foreign object damage (FOD) and cleanliness. If the compressor section is found dirty, perform an engine performance recovery wash, in accordance with the manufacturer's instructions. If other damage is discovered, consult the manufacturer's maintenance instructions to determine the extent of field repairs that are allowed or if the engine must be disassembled and repaired an overhaul facility.

Check the operation of all engine controls. Check power and fuel cutoff controls for full travel to the stops at the FCU (Also condition lever and propeller controls for turboprop engines).

Check for hang-up or distortion in any cam linkages.

Check the adjustment of the FCU especially if changes have been recently made during engine trimming operations.

2.   If the engine experiences an uncontrolled overspeed, check the FCU driveshaft for engagement and FCU bleed air reference lines for obstructions and leakage.

Check the FCU for contamination, and if detected, isolate the source and replace the FCU with a new or overhauled unit.

3.   The engine is slow to accelerate or surges during power application:

Check bleed air reference delivery tubes for leakage and obstructions.

Inspect the compressor for cleanliness and FOD damage. Perform a power recovery wash, as required.

Check bleed air valves for proper operation and the airframe auxiliary bleed air system for leakage.

Check the FCU fuel filters and screens for contamination. Check the airframe fuel source for contamination and fuel quality.

4.   When acceleration controls are installed on the FCU, attempt to make adjustments. If maximum adjustments do not obtain desired results, replace the FCU with a new or over-hauled unit.

5.   If the engine experiences a transient over temperature during operation (exhaust gas temperature or EGT, turbine inlet temperature or TIT, or interstage turbine temperature or ITT):

Perform a check of the engine temperature indicating instruments to determine the actual extent of the over temperature condition. Determine the maximum temperature value and the length of time of operation above maximum limits. (Consult the engine manufacturer's maintenance instructions to determine required actions. If the engine is allowed to remain in service, isolate the cause of the fault).

Check for excessive airframe accessory power loading. For example, excess generator loads or excessive bleed air requirements may cause transient engine over temperatures.

Perform a fiberoptic borescope inspection of the compressor inlet, blades, and guide vanes. Perform a power recovery wash if the compressor is found dirty or corroded. If corrosion is present, follow the manufacturer's instructions for continued operations.

Inspect the combustors and turbine sections for damage and distortion. Flow check the fuel nozzles and verify their alignment in the fuel manifold upon reinstallation.

6.   If excessive vibrations are indicated during normal operation:

Inspect the engine for loose mounting and the condition of all isolation dampners. For turboprop engines, check the propeller balance and the condition of the power section and gearbox.

Perform a visual inspection (with fiberoptic borescope as required) of the compressor and turbine sections for FOD damage.

Check the bleed air valves for proper operation.

Inspect the oil filter element or screen for signs of contamination and perform a spectrometric oil analysis.

Inspect the airframe fuel system for contamination and delivery pressures.

Check the engine driven fuel pump and FCU for proper fuel delivery.

Check the acceleration time of the engine and if outside of specifications inspect the FCU and fuel nozzles for contamination.

Inspect the compressor and turbine sections for FOD damage

7.   Low power or if all engine indicating parameters are showing low.

        Verify proper operation of the indicating systems.

        Check engine control linkages for proper operation and travel limits.

Check bleed air reference lines to the FCU and other engine accessories.

Attempt to perform engine trimming as specified, by the engine manufacturer. If unable to achieve desired results inspect the FCU and fuel nozzles for contamination and proper operation.

Unusual noises such as squealing or rubbing heard during operation:

Rotate the engine by hand to detect unusual rubbing at low rotational speeds. If squealing is only heard at high rpm inspect the compressor and turbine sections for evidence of blade tip to shroud rubbing.

Check all engine mounted accessories for proper operation.

Inspect the engine oil filter element or screen for evidence of contamination and perform a spectrometric oil analysis to check for wear of internal engine parts.

Cross-Bleed Starting

After one engine is started, the remaining engine(s) can be started with cross-bleed air, but this is not desirable under normal circumstances. It requires approximately 80 percent N2 speed to obtain the necessary air pressure, and this presents a problem of noise and jet blast on the ramp as well as a waste of fuel.

1.   To start #1 engine from cross-bleed air, open cross-bleed valves No. 1 and No.2.

2.   Open # 2 Augmenter Valve if air manifold pressure is low on a hot day or if customer bleed air is being used.

3.   Place Start Switch to "No.1" and follow procedure as for No.2 engine.

Air Starter Duty Cycle

The air starter has a prescribed duty cycle, typically:

1.   During engine starting, 5 minutes "On", 2 minutes "Off' to cool.

2.   During motor over checks, 5 minutes "On", due to additional loading 5 minutes "Off' to cool.

The reason for this restricted operating cycle time is that this unit contains a ring gear type reduction system which builds up friction heat very readily, and the low volume splash type wet sump oil system contained in the gear section has a limited cooling capability.

Other Starting Systems

Many other starting systems have been developed in the past for military and commercial engines. They are not in common use in either general or commercial aviation today.

1. High-Low Pressure Pneumatic Starter

An accessory gearbox mounted starter which is a type of air turbine starter that can utilize either conventional low pressure starting or a high pressure, 3,000 pounds per square inch, air source from an aircraft mounted storage bottle.

High pressure air starting (usually only on one engine) gives the aircraft a self-starting capability without the assistance of an auxiliary power unit or ground power unit

2. Cartridge-Pneumatic Starter

An accessory gearbox mounted starter which can use either an explosive solid propellant charge or a low pressure, high volume air source similar to the pneumatic (air turbine) starter. The charge is ignited electrically from the aircraft battery, giving the aircraft a self-starting capability without auxiliary power unit or ground power unit.

3. Fuel-Air Combustion Starter

An accessory gearbox mounted starter which utilizes a high pressure, 3,000 pounds per square inch, air source and a combustion process. It is very similar to a small gas turbine engine. Combustion is initiated electrically from the aircraft battery giving the aircraft a self-starting capability.

4. Turbine Impingement Starting

A low pressure, high volume air source of 45 pounds per square inch-gauge at 200 to 300 pounds per minute is directed onto the engine turbine wheel. The air source terminates after self accelerating speed is reached. No accessory is required in this system, only an inlet air port.

5. Hydraulic Starter

An accessory gearbox mounted hydraulic starter motor. It is driven by fluid from an auxiliary power unit mounted hydraulic pump, or a hand pump and accumulator arrangement.

## Chapter - 5

THRUST REVERSERS

Thrust Reverser or Propeller Reversing (Beta) System Checks -

Thrust reverser buckets rails, and actuators for condition and proper operation.

Inspect blocker doors for condition cracks damage and condition and paint

Inspect sealing door bearings for general condition a over heating

Inspect cascade vanes for nicks, dents over heating damage and security.

Propeller reversing components including pitch-stop limit components.

Reversing annunciator lights and indicating systems for proper operation.

Exterior Engine Condition Checks Including Inlet and Exhaust Areas

Exterior of all engine and accessory cases for cleanliness, cracks, distortion, and other defects.

Exhaust and inlets for cracks, wear, distortion, and other defects.

INTERIOR ENGINE CONDITION CHECKS

As discussed later in this section, check the interior condition of the hot and cold sections of the engine as prescribed by the inspection program, or when required as indicated by reductions in power output and/or acceleration performance.

Check accessory drive and power drive sections as required by the inspection program.

During overhaul all thrust reversal components must be dismantled, cleaned crack tested. If any cracks or parts, the same should be marked with the approved marker.

After crack test, the parts are inspected visually and some are dimensions. The parts which are dimensionally work sent for repair either in house repair or outside party repair. The cracked parts are inspected as per manual, if the cracks are within repairable limits, the cracks parts are repaired. If the cracks are beyond repairable limit, the part is scrapped.

The clam shell doors are checked for sealing after installation. The thrust reversal is to be checked for proper operation with the ground air supply for proper operation with the ground air supply. The accessories on thrust reversals has be functional tested. Some of which are no condition maintenance.

## Chapter -6

ENGINE MAINTENANCE

MONITORING CONDITION

Turbine Engine Troubleshooting

Turbine engine discrepancies can be either obvious or hidden. Hidden problems, when not detected and corrected, often cause major problems with continued engine operations. Hidden troubles are detected by continuously monitoring engine operating parameters, comparing them over time to detect subtle changes that indicate wear and transient problems. By analyzing engine trends and consulting the airframe and engine manufacturer's troubleshooting information, you can systematically isolate and correct defects before they cause major problems.

Trend Monitoring

One of the most useful procedures used to detect hidden and subtle problems with turbine engines is to perform continuous evaluations of engine performance parameters over time. A properly operating turbine engine produces fairly predictable and consistent compressor rpm, exhaust gas temperature, fuel flow, and power output for a given ambient pressure altitude and temperature. By recording parameter data during each flight, and analysing changes when they occur, problems can often be detected and corrected early in their development.

Performance monitoring of engines, commonly referred to as trend monitoring, is a technique used by aviation maintenance organizations and engine manufacturers to improve engine service life and reduce operating costs. The monitoring of an engine's performance and condition over a period of time provides a database for trend analysis: Trend analysis alerts an operator to deteriorating performance, providing an opportunity to take corrective action before substantial engine damage or failure occurs.

Proper trend monitoring begins when an engine is new or recently overhauled. Data collected during the initial operations establishes a baseline to compare to all subsequent operations data. After establishing an initial relationship between performance parameters, operating data is reviewed at regular intervals. Significant changes in the relationships between performance parameters may signal impending failures.

Data collection methods range from onboard computers to manual entries on paper forms. Regardless of the method used, conditions under which a certain parameter is measured should be consistent to be useful in a trend analysis. For example, EGT measurements and fuel flow indications should always be taken at the same power setting, and under the same ambient conditions. However, since ambient conditions often vary between flights, correction factors must be taken into account with data. If this is not done, it is possible to obtain data which skews the trend analysis and gives false indications of a problem.

Since it takes a great deal of experience to identify subtle trend variations, aircraft and engine manufacturers often provide trend monitoring services for their customers. In some cases, second party companies may also, provide these services. By using an FAA-approved trend monitoring program, engine operating times between overhaul or times and cycles between required inspections, may be extended. While trend monitoring improves safety, it also provides tremendous economical incentives.

Trend analysis information can be broken down into two broad categories; performance and mechanical. Typical performance parameters include information on EPR or torque meter (Np) readings, compressor N1 and N2 speeds, gas generator speed (Ng), fuel flow (Wf), and EGT or ITT. On the other hand, mechanical parameters typically include instrument readings for oil pressure, oil temperature, oil quantity, vibration, oil pressure warning lights, and bypass lights. Accurately interpreting a trend analysis requires the ability to discern small shifts in operating parameters on one or more gauges and to accurately compare the information to base line data.

Troubleshooting Resources

Instructions for troubleshooting gas turbine powerplants are contained in the aircraft and engine manufacturer's maintenance manuals. The information contained in the airframe manual provides operational considerations in relation to the airframe systems and general powerplant troubleshooting guidance. For additional, and often more specific engine troubleshooting information, consult the powerplant manufacturer's manuals.

Standard engine instrument readings are used with troubleshooting guides to provide clues as to the cause of a given engine malfunction. In addition, some aircraft are equipped with built-in test equipment, or BITE test systems. BITE systems consist of sensors, transducers, and computer monitoring devices, which detect and record engine data such as vibration levels, temperatures, and pressures. A typical BITE test requires you to make entries on a keypad in the cockpit or at a remote terminal in order to receive engine data. However, even this sophisticated equipment can only provide you with the symptoms of a problem, leaving you to determine the actual problem.

A typical turbine engine maintenance manual provides one or more troubleshooting tables or flow diagram charts to aid in pinpointing the cause of common malfunctions. However, it is not possible for these to identify all malfunctions. Instead, troubleshooting tables and flow charts provide you with a starting point in the troubleshooting process. This, combined with a thorough knowledge of the engine systems, logical reasoning, and experience provides you with the information necessary to diagnose and correct complicated or intermittent malfunctions.

Most powerplant manufacturers group engine troubles into categories similar to the following:

Starting and shutdown faults Operational faults

Performance faults and engine trend monitoring (ETIVI) shifts

Lubrication faults and oil contamination

To isolate a fault, it is necessary to have reports of previous problems and any actions that may have been taken to correct them. You should check the probable source of the trouble, by use of diagnostic tests, until the defect has been isolated. As always, use a systematic sequence to isolate the fault to prevent making wrong and costly assumptions as to the cause of a problem.

INDICATED MALFUNCTION	POSSIBLE CAUSE	SUGGESTED ACTION
Engine has low RPM, exhaust gas temperature, and fuel flow when set to expected engine pressure ratio	Engine pressure ratio indication has high reading error	Check inlet pressure line from probe to transmitter for leaks. Check engine pressure ratio transmitter and indicator for accuracy
Engine has high RPM, exhaust gas temperature and fuel flow when set to expected engine pressure ratio	Engine pressure ratio indication has low reading error due to Misaligned or cracked turbine discharge probe. Leak in turbine discharge pressure line from probe to transmitter. Inaccurate engine pressure ratio transmitter or indicator.	Check probe conditions Pressure test turbine discharge pressure line for leaks. Check engine pressure ratio transmitter and indicator for accuracy
NOTE: Engine with damage in turbine section may have tendency to hang up during starting.	If only exhaust gas temperature is high, other parameters normal, the problem may be thermocouple leads or instrument	.... gas temperature instrumentation
Engine vibrates throughout RPM range, but indicated amplitude reduces as RPM reduced.	Turbine damage	Check turbine as outlined in proceeding item
Engine vibrates at high RPM and fuel flow when compared to constant engine pressure ratio.	Damage to compressor section	Check compressor section for damage
Engine vibrates throughout RPM range, but is more pronounced in cruise or idle PRM range.	Engine mounted accessory such as constant speed drive, generator, hydraulic pump etc.	Check each component in turn
No change in power setting parameters, but oil temperature high.	Engine main bearings	Check scavenge oil filters and magnetic plugs
Engine has higher than normal exhaust gas temperature during take-off climb and cruise RPM and fuel flow higher than normal.	Engine bleed air valve malfunction Turbine discharge pressure probe or line to transmitter leaking	Check operation of bleed valve check condition of probe and pressure line to transmitter.
Engine has high exhaust gas temperature at target engine pressure ratio for takeoff.	Engine out of trim	Check engine with jetcal, Retrim as desired
Engine rumbles during starting and at low power cruise conditions.	Pressurizing and drain valve malfunction Cracked air duct Fuel control malfunction	Replace pressurizing and drain valves Repair or replace duct Replace fuel control
Engine RPM hangs up during starting	Subzero ambient temperature Compressor section damage Turbine section damage	If hang up is due to low ambient temperature, engine usually can be started by turning on fuel booster pump or by positioning start lever to run earlier in the starting cycle. Check compressor for damage inspect turbine for damage.
High oil temperature	Scavenge pump failure Fuel heater malfunction	Check lubricating system and scavenge pumps. Replace fuel heater.
High oil consumption	Scavenge pump failure High sump pressure Gearbox seal leakage	Check scavenge pumps. Check sump pressure as outlined in manufacturer's maintenance manual Check gearbox seat by pressurizing overboard vent.
Overboard oil loss	Can be caused by high airflow through the tank, foaming oil, or unusual amounts of oil returned to the tank through the vent system	Check oil for foaming vacuum check sumps check scavenge pumps.

Figure 14-24 Turbine engine troubleshooting guides provide suggestions to assist you in determining the cause of a malfunction

PERFORMANCE FAULTS AND ENGINE TREND MONITORING (ETM) SHIFTS

As previously discussed, most turbine-powered aircraft engine performance is continuously evaluated through a trend monitoring program. By performing a trend analysis, corrective actions can be taken to correct an unsafe condition as soon as the engine's operating parameters deviate from their established baseline.

It is not uncommon for turbine engines to display slight parameter variations and degradation in performance as a factor of normal wear. Significant changes indicate abnormal conditions that must be isolated to assure maximum engine efficiency and safety. Variations in fuel flow (Wf), temperatures, and power (N1, EPR or torquemeter Np) are the primary parameters used to evaluate the engine's performance, but other parameters such as oil pressure and chip detector warning systems are also used to indicate mechanical irregularities.

Common performance faults include the following symptoms along with their possible causes:

Rapid shift in engine temperature monitoring parameter that is not accompanied by shifts in fuel flow, compressor speed, or power output:

An indicator or sensing system fault almost always causes rapid temperature shifts without other parameters being affected. Check the condition and operation of temperature probes and bus bars. Perform an operational check of the temperature sensing system using a Jetcal analyzer, or similar calibration equipment.

For multi-engine aircraft, when both engines have a similar rapid temperature shift, check the airframe outside air temperature (OAT) system, and altitude or airspeed indicating systems since malfunctions may cause erroneous data reports from the flight crew.

A slight shift of engine temperature parameters without other accompanying parameter shifts may indicate fuel nozzle contamination. Remove the fuel nozzles for flow checks and verify nozzle alignment in the fuel manifold upon reinstallation.

Changes in a single operating parameter without corresponding changes to other operating parameters, or changes to all operating parameters in a proportionate amount.

Again, if fuel flow shows a rapid change without an engine temperature or compressor speed change, this is usually indicative of an instrument system fault. Perform a Jetcal analysis to check each instrument indicating system.

If all parameters indicate a decrease in a proportionate amount, it often indicates that the engine power sensing, or airframe OAT, airspeed, or pressure altitude instruments are at fault. For example, if the EPR or torque meter is in error, the flight crew will set power in accordance with the erroneous indication. This will always be reflected in changes to all other operating parameters in proportionate amounts.

Exhaust temperature, compressor speed, and fuel flow all show an increase beyond baseline when desired power settings are established:

Perform a calibration check of all engine instruments and indicating systems.

Check for engine inlet obstructions, distortion, and FOD damage to the compressor.

Perform a power recovery wash if the compressor is found dirty or corroded. However, this is generally not considered to be the fault when sudden parameter shifts occur, but may be useful to partially regain performance losses.

Check the operation and condition of all bleed air valves and airframe auxiliary bleed air systems. Airframe systems can be isolated from the engine by the installation of blanking plates, as previously discussed.

Check for any seal damage or other sources of hot air bring ingested back into the engine intake. For ground operations, this may be caused by the use of reverse thrust or by high winds blowing into the tailpipe or carrying hot exhaust gases toward the engine intake.

Determine if any hot starts may have been recently encountered, which may explain the rapid parameter shift. If evidence suggests that a hot start may have been encountered, perform a hot section inspection to determine the engine's internal component integrity. Consult the engine manufacturer's hot start inspection procedures and over temperature limitations.

Engine temperature and fuel flow increase while compressor speeds decrease or remain unchanged:

Inspect the turbine section for damage to exhaust nozzles, turbine blades and turbine disk.

Check the pressurization and dump valve (purge valve) for proper operation.

Check for bleed air leakage and. for proper bleed air valve operation.

Inspect the compressor section for damage and cleanliness. Perform a power recovery wash if the compressor is found dirty or corroded.

HUMAN FACTORS DURING INSPECTIONS

Through an analysis of aircraft accidents and incidents that occurred during the past four decades, it has been determined that the rate of aircraft accidents and incidents due to structural and mechanical failures has decreased. On the other hand, the accident and incident rates that were attributed to human error increased, negating many of the safety improvements in aircraft design and mechanical reliability. Further analysis revealed that, in many cases, policies and procedures were in place that, if properly followed, would have prevented many accidents and incidents that were attributed to maintenance personnel errors. Accidents and incidents that occur because of human error or over-sight are commonly referred to as human factor events.

The aircraft industry has recognized the potential for human errors in flight operations since the inception of aviation. For example, preflight inspections have always been performed to help identify and correct problems before flight. During World War II, human factor events were further reduced by the adoption of checklists and consistent policies and procedures for conducting aircraft inspections. Even with these and many other improvements, today, maintenance related human factor events still occur.

Some areas where human errors occur during aircraft inspections include the following:

Failure to properly follow regulatory and manufacturer's inspection instructions including incomplete work due to missed steps or failure to follow maintenance instructions in the correct sequence.

Failure to perform follow-up maintenance activities with an effective secondary inspection by an experienced supervisor or someone with equal or better qualifications than the person who performed the original inspection.

Failure to identify defects due to improper or inadequate inspection procedures or complacency on the part of individuals during the inspection or maintenance check.

Inappropriate or inadequate use of special tools when conducting inspections.

Deviation from established and approved inspection methods and procedures.

Of course there are many other areas where human error can enter into aircraft maintenance and inspection operations. In many cases, repair facilities have sufficient policies and procedures in place to minimize human factor events from occurring However, if policies and procedures are not properly followed, or if discrepancies in policies and procedures are not identified then the potential for maintenance related human factor accidents and incidents increases dramatically.

If you become employed as a technician with a maintenance repair facility, you are required to follow your company's policies and procedures when conducting aircraft inspections. To ensure maximum safety it is important that you abide by these policies and procedures, but you should also be alert to identifying any deficiencies in your company's inspection programs. When you identify policies that need refinement, or determine that inappropriate procedures may be- affecting productivity, bring them to the attention of your supervisor, or other upper management personnel. Typically, there is room for improvement in any policies and procedures, but changes must be thoroughly reviewed and approved by all affected departments within the repair facility, and ultimately, approved by the FAA or other regulating agency.

Inspection Programs

Turbine engines are installed on many types of aircraft. Helicopter manufacturers use turboshaft engines in their aircraft because of the engine's compact size, light weight, and ability to operate in confined spaces with little exterior cooling airflow. The high reliability of turbine engines has also led a large number of manufacturers to develop turbine-powered, single-engine airplanes that offer an economical alternative to multi-engine designs. However, multi-engine airplanes constitute the majority of turbine-powered aircraft in use today.

A continuous airworthiness inspection program that is part of a continuous airworthiness maintenance program (CAMP).

An approved aircraft inspection program (AAIP).

An inspection program recommended by the aircraft manufacturer.

Any other inspection program established by the registered owner or operator of the aircraft and approved for use by the FAA or other regulating agency.

Each of these programs provides adequate inspection intervals and detail to reasonably guarantee the airworthiness of the aircraft between inspection events.

REPAIR FACILITY INSPECTION PROCEDURES

Individuals that own their own turbine aircraft often elect to have required inspections performed by FAA certified repair stations that specialize in maintenance on their specific aircraft make and model. On the other hand, fleet aircraft owners and air carriers often establish their own maintenance shops to maintain their aircraft. In either case, FAA-approved repair facilities must establish policies and procedures for conducting aircraft inspections. When employed as an aircraft engineer for a repair facility, they must be familiar with company's inspection procedures to understand the specific responsibilities, as well as the responsibilities of others within the organization.

Procedure Manuals

Repair facilities such as repair stations and certificated air carriers must provide the FAA with procedures manuals that designate and define the responsibilities of personnel assigned to perform aircraft inspection activities.

Due to the complexity of aircraft inspections on turbine-engine aircraft, repair facilities are often organized into departments that perform specific inspection tasks. For example, it is common for maintenance facilities to have a department that tracks all required aircraft records. Employees that are assigned duties within this department are typically responsible for reviewing all aircraft records before an inspection begins to determine the following:

The status of life-limited components based upon hours of operation, flight cycle times, or other criteria.

Records of major alterations and repairs including changes to weight and balance reports and the aircraft's equipment.

A determination as to what items must be inspected to meet the requirements of the particular phase, or cycle, of the aircraft's inspection program.

A common practice used by many facilities is to issue work cards to their technicians that provide instructions for performing each inspection task. Once the task is complete, the technician signs the work card and has an inspector perform a secondary inspection of the completed task. When it has been determined that the task was done properly, the inspector also signs the work card. The technician then returns the card to the records department where it is tracked and signed off as complete. in accordance with the company's inspection program procedures. The practice of having all inspection items inspected by a technician, and a final inspection conducted by an inspector, ensures that more than one person conducts the inspection operation. From a human factors standpoint, this technique provides improved safety by helping to identify items that may have been missed 'during the initial inspection, or procedures that may have been improperly performed.

With regard to turbine engine inspections, the complexity of each inspection task depends on the particular phase, or cycle, of the inspection that is being done. The task may consist of reasonably minor items, such as checking and servicing the oil, or more extensive activities, such as performing an interior inspection of the engine's hot section. Regardless of the complexity, the Work card provides detailed instructions on where to find pertinent information in the aircraft or engine manufacturer's maintenance manuals regarding the actions to take to complete the task.

Specific details of items to be checked during a particular inspection are contained in the airframe manufacturer's maintenance manual. For example, manufacturers that use ATA Specification 100 coded manuals provide inspection program information in Chapter 5. Often, however, manufacturers convey other inspection information through Service Bulletins, or Service Communique. As always, it is important to remember that when using any maintenance information, you must ensure that it is the most current available from the manufacturer. The repair shop should be equipped with all manuals like overhaul manual, maintenance manual, IPC and Standard practice manual.

Inspection Tools

To aid engineers in performing turbine engine inspections, many engines are equipped with openings, or ports, that allow to inspect the inside of the engine without disassembly. Instead of inspection ports, some engines are constructed in a way that allows o remove the fuel nozzles, igniter plugs, and other components to gain access to the interior of the engine. Some common tools used to inspect the inside of a gas turbine engine through access openings are the borescope, fiberscope, and electronic imaging.

A borescope is an internal viewing device which allows you to visually inspect areas inside a turbine engine without major component disassembly. A borescope may be compared to a small periscope with an eyepiece at one end and a strong light, mirror, and lens at the other end. A conducting cord connects the probe to a control device for adjusting light intensity and lens magnification or focusing.

A fibre optic borescope, or fiberscope is similar to a standard borescope, but has a flexible articulating probe that can be bent around corners. This allows you to view areas deep inside an assembly that previously required disassembly to inspect. In a typical fiberscope, a bundle of optical glass fibres transmit light from a light source to the probe end of the scope, and also serve to carry back the viewed image to the lens for viewing. The probe is then inserted into the component or structure being inspected. The maximum length available for fiberscopes is generally around four feet.

Special attachments allow mounting a camera to the fiberscope, thereby allowing you to view through the scope from a remote monitor. These attachments usually allow you to attach recording devices, such as a video recorder, to maintain a record of the inspection. Some scopes also provide a hollow chamber that allows you to spray inspecting dyes, such as dye penetrant liquids, toward a part requiring a more thorough inspection. In addition, the hollow channel may also be used for inserting other special accessories, such as scribes and extracting tools. These allow you to manipulate parts within the engine without disassembly.

Special training is also required before conducting fiberscope inspection on turbine engines because of the difficulty in determining the dimensions and extent of damage to components when damage is found. For example, the magnification of images can cause relatively minor defects to appear substantially large. A technique that experienced inspectors sometimes use is to securely attach a feeler gauge or safety wire of known thickness to the fiberscope in such a way that it is visible through the eyepiece. Once the gauge is in position near the defect, a comparison can he made to determine the actual size of the defect.

Electronic imaging in a relatively new technique which provides sharp, colour images on a video monitor. Since the image is in colour and can be magnified, it is much easier for inspectors to differentiate between an actual defect and an unclear image.

Fibre optic technology and light emitting diodes (LEDs) are utilized in many video imaging systems. Computer enhancement provided by this type of inspection equipment is particularly useful when videotaped because it allows several inspectors in different locations to consult on a given image. In addition, recordings allow evaluations of progression of damage over time through multiple inspections.

Inspection Designations

Aircraft and engine manufacturers designate turbine engine inspection tasks to help classify and identify specific inspection activities. These designations are often grouped into routine or non-routine inspection classifications.

Routine Inspections

Routine, inspections are those that are mandated by an approved inspection schedule or Federal regulations. All routine inspections are performed periodically at intervals specified by FAA-approved company or airline operations manuals. Examples of routine inspections include preflight inspections, airworthiness inspection programs, cold section inspections, and hot section inspections.

In some cases, routine engine inspections may be carried out as a function of line maintenance. At other times, the inspections are conducted within a repair facility when the rest of the aircraft is undergoing a detailed phase inspection.

The routine inspections that are part of a continuous inspection program are often given a specific name. Some examples of terms used by air carriers to define routine inspections include "number 1, service," "number 2 service," "A" check and "E" check. The content and inspections required for each of these checks is defined in the inspection procedure Manual.

INSPECTION

The following paragraphs describe the inspections which are normally required by the engine manufacturer and specified in the appropriate Maintenance Schedule.

During schedule inspection there are various types of inspection are carried out to maintain an engine in airworthy condition. The various types of inspection are

1.   Pre-flight inspection

2.   Daily inspection

3.   Transit inspection

4.   Base terminal inspection

5.   Layover Inspection

6.   Hot section inspection

7.   Heavy maintenance inspection

8.   Overhaul inspection

The visual inspection is carried out on the following parts.

General Precautions

All turbine engines are susceptible to damage through the ingestion of foreign bodies. The axial flow type of compressor is particularly vulnerable to this type of damage and extreme care is necessary when carrying out an inspection of any part of the engine or adjacent structure to ensure that loose articles are not left where they may subsequently be drawn into the compressor. Clothing, buttons, caps, belts, pencils and tools are all items which, if left in intakes or engine cowlings, could cause damage necessitating removal of the engine. Intake and jet pipe covers should always be used when an aircraft is parked for any length of time and the intake area should be inspected for foreign objects or debris immediately prior to starting the engine. During cold section inspection the following Parts are to be inspected.

## 1.	Air Intake

The air intake passage leading to the compressor is designed to permit the required flow of air into the engine under all conditions of flight. Any damage or blockage in the air intake could affect engine performance. All panels, fairings, fasteners, bolts and locking devices should be inspected for security of attachment, flush fitting or damage. Unserviceable items should be 'repaired. or renewed.

Intake lips should be inspected for dents or other damage which could affect the airflow. Air bleeds or drain holes should be clear. Pressure sensing probes, where fitted, should be inspected for security, cracks and corrosion.

Electrically heated anti-icing mats located in the air intake should be inspected for adhesion and damage. Controls for these heater mats may be automatic, continuous or cyclic and the manufacturer's manual should be consulted for details of a particular system before carrying out a functional or electrical resistance test.

## 2.	Compressors

The compressor is an accurately manufactured and balanced component which may become damaged by hail, ice or foreign bodies. Inspection of the guide vanes and as many rows of stator and rotor blades as possible should be carried out with the aid of a mirror and strong spot light.

### a.	Superficial Damage

Damage to. aluminium, steel or titanium blades in the form of small dents, may be regarded as acceptable provided that it cannot result in the propagation of cracks and is within the limits defined in the manufacturer's manual.

### b.	Impact Damage

Impact damage caused by the entry of small metal parts can result in rejection of the engine as the damage may extend throughout the compressor. If thread or hexagon impressions are visible on the stator or rotor blades, the engine should normally be removed for strip examination.

### c.	Reparable Damage

Damage caused to the initial compressor stages by small stones or grit may be blended out within specified limits. A clean cloth should be placed round the blade being repaired to catch any swarf or filings removed during the blending process. File marks on blades should be removed with fine emery cloth and the blades inspected for cracks by a suitable penetrant-dye process immediately after the work and at regular intervals during the remaining life of the engine. Renewal of the anti-corrosive treatment may also be required.

Diagnosis of damage to later compressor stages may only be possible from the results of engine performance and running checks Where limited engine stripping to facilitate inspection is permitted, procedures. will be detailed in the appropriate Maintenance Manual.

3. Compressor Casing

The compressor casing should be examined for signs of damage, gas or fluid, leaks and the components for security of attachment. Any corrosion should be removed, the affected areas retreated and damaged paint renewed as necessary. NOTE: Compressor casings are often manufactured from magnesium alloy which requires special anti-corrosive treatment, details of which will be given in the manufacturer's manual.

(a)   Certain components attached to the casing may be fitted with 'witness' drains and these should be examined to ensure that the permitted leakage is not exceeded.

4. Centrifugal Compressor Inspection

Inspect the impeller for nicks, dents, and cracks, using the following criteria:

a.   Critical Area. No cracks or nicks allowed. Smooth dents are permitted, provided they do not exceed 0.030 inch in diameter and/or 0.010 inch in depth.

b.   Leading Edge. No cracks allowed. Nicks or dents are permitted, provided they do not exceed 0.100 inch in depth and 0.300 inch in length after blend repair. Distance between repairs must be equal to or greater than length of longest repair.

c.   Trailing Edge. No cracks are allowed. Nicks or dents are permitted, provided they do not exceed 0.060 inch in depth or 0.300 inch in length after repair.

Distance between repairs must be equal to or greater than length of longest repair.

d.   Blade Tips. No cracks are allowed. Six nicks or dents are permitted] provided they do not measure 0.060 inch in depth or 0.300 inch in length after repair. Distance between repaired areas must be at least 3/8 inch.

e.   Airfoil sides of blades). No cracks are allowed. Nicks and dents are permitted, provided they do not exceed 0.030 inch in depth or 0.350 inch in length after repair. Distance between repaired areas, regardless of location, must be at least 1/2 inch.

5. Axial Flow Compressor

Inspect axial flow compressor blades for nicks, dents, cracks and tears. Inspect and ensure that the compressor disks are possessed sufficient lift.

HOT SECTION PARTS INSPECTION (HSI)

During hot section inspection the following parts are to be inspected.

1.   Combustion Section

Inspection of the internal features of the combustion chamber is not always possible in service, unless the engine is equipped for this Where individual flame tubes are fitted the manufacturer may recommend their removal at specified intervals in order to carry out a detailed inspection for cracks or other heat damage Removal of the -flame tubes will permit an inspection of the nozzle guide vanes and high pressure turbine blades. In other cases the manufacturer often specifies an overhaual of the 'hot' section of the engine which is considered to be equivalent to the top overhaul of a piston engine. These inspections are often possible in situ, but, as radiographic or radio isotope crack detection methods are usually specified, it is often more convenient to remove the engine for inspection at a suitably equipped workshop.

Inspection during service is normally confined to an external examination of the combustion section as follows:

a.   Heat shields and lagging should be inspected for security, signs of burning or gas leakage. Any signs of overheating are an indication that internal damage may have occurred and further examination may be required.

b.   Fuel injectors should be inspected for security, damage or fuel leaks.

c.   The ignition system wiring should be inspected for signs of chafing, security and fluid contamination.

d.   Drain pipes should be inspected for security and must be free from blockage. A blocked combustion chamber drain pipe constitutes a potential fire hazard in certain circumstances. On some aircraft the drains are led to a common collector box which is emptied by suction during flight; these may require manual draining following a failure to start the engine.

2.   Turbine and Exhaust Section

A strong spotlight and appropriate viewing equipment are necessary—when inspecting the turbine and exhaust section All components should be carefully checked for damage, cracks or signs of metallic deposits.

Damage to turbine blades, 'other than slight pitting, is not usually acceptable.

Damage to other components, including small cracks in jet pipes, is often permitted. Metallic deposits most likely to be found are aluminium, showing as a dull white or silver splatter, and titanium, which may be in the form of bright blue or golden speckles. Heavy deposits will necessitate removal of the engine for further examination but light deposits should be assessed in relation to engine performance before resorting to engine removal.

The jet pipes of turbo-prop engines are not so highly stressed as those of turbo-jet engines because there is little residual thrust left in the exhaust gases after the turbines have taken the power necessary to drive the propeller. Cracks or holes in these jet pipes may be repaired by patching with a similar material of the same gauge, electric resistance welding is usually specified to limit distortion. The manufacturer's manual should be consulted regarding the extent of repairs permitted and any subsequent limitation on time in service or inspection frequency.

Thrust reversers are often employed in large aircraft and should be subjected to an inspection similar to that carried out on jet pipes. It is particularly important that the doors or buckets used in thrust reversers should be flush with the jet pipe when retracted. Any projection will result in a 'hot spot', which will lead to distortion and cracking. Thrust reverser operating mechanisms should be checked for correct operation in accordance with the appropriate Maintenance Manual.

3.   Buried Engines

Additional checks should be carried out on engines buried within the airframe structure as follows:

(a)   Long jet pipes are often employed with buried engines and are lagged to prevent the transfer of heat to the surrounding structure. An inspection should be made to ensure that there are no gas leaks at the joint with the engine or cracks in the pipe itself, indicated by burning of the lagging material.

b)   Because of its length the jet pipe is suspended in such away as to allow for considerable axial expansion. The fixed attachments should be checked for security and the expansion links for freedom of movement.

(c)   The surrounding airframe structure should be inspected for signs of excessive heat such as discoloration or blistered paint. There should be adequate it' clearance from the jet pipe to permit the passage of cooling air.

Engine Mountings. Engine mounting structures are designed to allow for engine expansion and usually consist of a main trunnion to transmit engine thrust to the airframe and a secondary support to steady the engine. Turbo-prop engines are normally mounted in a tubular cage similar to that used on radial piston
engines.

Turbo- jet engine mountings should be inspected for security, cracks and corrosion. Cracks are not acceptable and any corrosion must be removed and the component re-protected. The mountings are usually subjected to a detailed crack detection test whenever the engine is removed.

Tubular mounting structures should be inspected for dents, bowing, cracks or corrosion and a dimension check carried out whenever the engine is removed. Repairs are usually permitted within specified limits and the procedures in the appropriate Maintenance Manual should be followed. When engines ate subjected to shock loading as a result of a heavy landing or damage to a propeller, the engine mountings should be critically examined for cracks, bending, pulled rivets or other signs of distortion (buckled cowlings are a good indication of damaged mounting structures). The torque loading of bolts held in tension should also be checked; a low value indicates that the bolt has stretched and should be renewed. The mounting should be given a complete dimensional check and tested for cracks by the magnetic-flaw method following engine removal.

Internal Inspection

In addition to the checks detailed above for specific components some modern engines are provided with access holes at strategic positions through which an endoscope inspection can be carried out. By this means certain internal features such as combustion chambers and compressor or turbine blades can be examined without recourse to engine removal and stripping.

Limitations

The power obtainable from a gas turbine engine is limited by the ability of the materials used in its manufacture to withstand the high centrifugal forces and high gas temperatures developed within the engine. The life of components in the 'hot' sections of the engine, i.e. combustion chambers, turbines and jet pipe, is also influenced by the number of temperature cycles to which they are subjected. It is mainly the construction of the turbines which decides the operating speeds and temperatures of the engine and although operation within these limits is often mechanically controlled, care must be taken to ensure that they are not exceeded either during ground running or in flight.

Indicators

Various conditions of the turbine engine must be known to enable satisfactory operation within limitations, and indicators are fitted to the aircraft instrument panel for this purpose. Their functions are explained in the following paragraphs.

Engine Speed

The speeds of the rotating assemblies must be known so that the centrifugal forces acting on the compressor and turbine rotors may be kept within safe limits.

Propeller speed is also indicated when appropriate. Indicators are usually electrically operated and are calibrated to show revolutions per minute or a percentage of maximum rotational, speed.

Turbine Gas Temperature

Turbine gas temperature is a critical variable of engine operation but because of the high temperatures at the combustion chamber exit it is impractical to measure the actual temperature of the gas impinging on the turbine blades. The temperature drop across turbines is a known quantity, however, and temperature probes are often placed either in the stator blades aft of the first or second stage turbine or in the jet pipe. Indicated gas temperature limitations will thus differ considerably between engines and a low gas temperature limitation must not be taken as being less significant than a high one. Normally, a number of thermocouples are arranged radially around the engine with the mean temperature shown on a single indicator. This is known as the Exhaust Gas Temperature (EGT), Turbine Gas Temperature (TGT) or Jet Pipe Temperature (JPT) gauge, depending on the position of the probes.

Power

The power or thrust developed by an engine is not always directly proportional to gas temperature and rotational speed. Other factors such as ambient air pressure and temperature and forward speed also affect the power produced. 'Power' is measured in the form of thrust or torque and calibrated on an indicator which also acts as a power deficiency warning.

(a)   The thrust of a turbo-jet engine is measured from the jet exhaust pressure. A commonly used indicator is the engine pressure ratio (EPR) gauge which measures the ratio of exhaust pressure to air intake pressure. Alternatively, a simple pressure gauge measuring jet pipe pressure is used. An accurate figure for comparison purposes is obtained by correcting for ambient temperature. In the case of a turbo-fan engine, the fan and exhaust pressures are measured.

(b)   On turbo-prop engines the torque produced at the propeller shaft is measured because jet thrust is only a small proportion of the engine power. One particular type of torquemeter measures the oil pressure required to oppose the axial thrust of the helical teeth on the reduction gear, the gauge being calibrated in lbf/ in^2.

Engine Vibrations

The repairs described in this chapter for all rotating airfoils ensures that strength, aerodynamics, and balance are maintained. The original design of the engine provides for little or no inherent imbalance in the engine by engineering techniques to eliminate vibration, resonance, and harmonics.

These terms are defined as follows:

Natural frequency of vibration - the RPM at which a rotating object will vibrate Engine parts are rotated so that natural frequency speeds are never reached. Improper repairs however, can lower the natural frequency into the operating RPM range.

Resonance - occurs when two mechanical systems, closely mounted, have the same natural frequency of vibration. If one malfunctions and reaches its natural frequency, the other will also vibrate along with the first, failing both systems.

Harmonics - severe vibrations that occur at one-times the magnitude of vibration at the natural frequency and, if unattended, two-times the natural frequency, and so on until the part fails.

Other Indicators

Gauges, warning lights or magnetic indicators are fitted to show fuel and oil system functions. Some multi-engine aircraft are fitted with a synchroscope to assist in synchronising engine speeds during flight.

Foreign Object Damage

Much of the damage to the compressor section encountered on the flight line arises from foreign matter being drawn into the engine inlet Damage to compressor blades results in a compressor geometry change which can cause malfunctions such as performance deterioration, compressor stalls, and even engine failure Foreign object damage can also be caused by objects dropped into the engine gas - path or internal parts during maintenance.

Foreign object damage (F.O.D.) prevention is a concern of all flight line personnel. Sometimes the most harmless looking piece of debris can cause thousands of dollars in maintenance costs to the aircraft owner. The following is a list of suggested F.O.D. prevention methods hr managers, technicians and pilots. Ensure that:

1. Maintenance personnel keep ramp and hangar areas clean:

2. Other flight line personnel are appraised of the importance of cleanliness in work areas;

3. All personnel keep articles of clothing and materials in .pockets secured when working on operating aircraft;

4. Engine operators check inlets for foreign objects before engine run-up and that they avoid run-ups and taxiing into exhaust blasts of other operating aircraft;

5. Everyone who maintains jet aircraft keeps the inlet and exhaust covers in place when the engine is static (not operating) to prevent contamination and windmilling.

NOTE: Damage incurred in the gas path from material failure of aircraft or engine parts is normally termed Domestic Object Damage (D.O.D.) rather than foreign object damage.

Erosion

Gas path erosion occurs from ingestion of sand, dirt, dust, and other fire airborne contaminants. This ingestion affects both the compressor and the turbine sections. The abrasive effect of repeated ingestion can wear through the surface coating and even into the base metals of the fan, the compressor blades and vanes. It can even cause similar damage to the turbine before leaving the engine via the exhaust.

Designers of modem aircraft better understand this problem today and try to engineer the slip-streams around the aircraft to carry contaminants around rather than into the inlets. However, many older aircraft have ingestion problems. Also, some have been reconfigured from narrow nacelles to wide, high bypass fan engine nacelles and have very low ground clearance. These aircraft, especially, are experiencing performance loss and an increase in maintenance and fuel costs due to the effects of erosion on compressor and turbine parts.

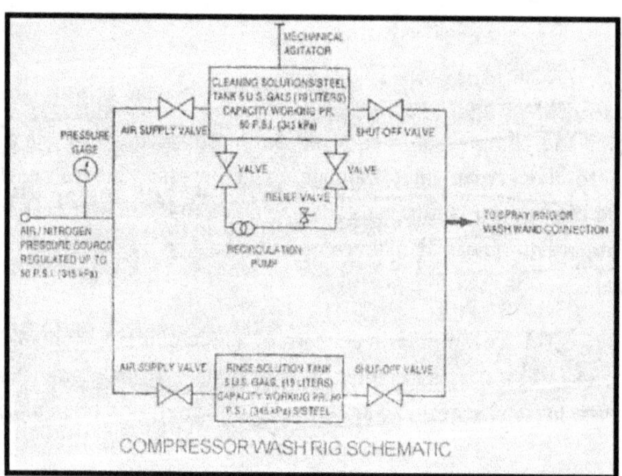

Compressor Field Cleaning

Accumulation of contaminants in the compressor reduces aerodynamic efficiency of the blades, thereby reducing engine performance. Contaminants, mainly salt airborne pollutants from smokestacks and a agricultural chemicals, all pass through the engine and build up over time on internal surfaces.

Two common methods for removing dirt, salt, and corrosion deposits are a fluid wash and an abrasive grit blast. Before field cleaning, it may be necessary to blank: off certain sensing and bleed ports to prevent contamination or blockage.

a.   **Fluid Cleaning Procedure**

The fluid cleaning procedure is easily accomplished by first spraying an emulsion type surface cleaner and then applying a rinse solution into the compressor. This is done while the engine is being motored over by the starter or during low speed operation. It cannot be over stressed that the wash procedure must be performed in strict accordance with the instructions set forth in the manufacturer's maintenance manual. There are two methods of performing compressor washes.

1.   While motoring the engine with starter only .

2.   While running the engine.

Depending on the nature of the operating environment and the type of deposits in the engine gas path, either of the two wash methods can be used to remove salt or dirt and other baked on deposits which accumulate over a period of time and cause engine performance deterioration.

When the water wash is performed solely to remove salt deposits, the compressor wash is known as a "desalination wash".

When the solution wash is performed solery to remove baked on deposits to improve engine performance, the compressor wash is known as a "performance recovery" wash.

The motoring wash is carried out at engine speeds of 14% to 25% and with the cleaning mixture injected at a pressure of 30 to 50 pounds per square inch.

The running wash is carried out at engine speeds of approximately 60% with the cleaning mixture and rinsing solution injected at a pressure of 15 to 20 pounds per square inch gage.

The fluid wash method also cleans the turbine area.

Turbine sulfidation (sulfur deposits from burning fuel which collect on turbine components) causes surface damage over time, and the frequent use of surface cleaning solvents in a motoring wash procedure has been found to be an added benefit in terms of extended engine service life for some engines.

The timely use of fresh water rinsing, where prescribed, to remove salt deposits, and use of inlet and exhaust plugs will greatly reduce the need for these heavy cleaning procedures. Some manufacturers of small engines have authorized fogging of the compressor with anti-corrosive fluids after the wash procedure to slow the contamination process.

b.   Abrasive Gift Process

A second more vigorous method of cleaning is to inject an abrasive grit (one popular material ground walnut shells or apricot pits trade name Carbo-blast) into an engine operating at selected power settings. The amount of material and the operation procedure is prescribed by the manufacturer for each particular engine. The greater capability of this procedure over the solvent and water methods allows the time interval between cleaning to be longer, but because the cleaning grit is mostly burned up in combustion, the agent does not clean the turbine vanes and blades as effectively as the fluid wash.

Scheduled Line Maintenance

Scheduled line maintenance includes inspections, such as the 100-hour, annual, continuous and progressive, which are frequent tasks performed by the maintenance technician. FAR Parts 43, 65, and 91 describe the scope of these inspections, and the manufacturer publishes the specific inspection procedures for the particular engine being inspected.

A typical 100 hour inspection on the turbine engine in a small business jet would include, but not be limited to, the following items:

1.   An oil change every third inspection (300 hours).

2.   The oil filter inspected and then cleaned or replaced.

3.   An analysis of the oil filter and chip detector debris.

4.   The oil analyzed by means of a spectrometer check.

5.   The igniter plugs visually inspected and opera-tionally checked.

6.   The fuel pump filter(s) cleaned or replaced.

7.   A visual inspection of the engine inlet and exhaust, to check for possible FOD and turbine distress.

8.   A visual' inspection of the engine's exterior, checking for leaks and the condition of the lines and electrical leads.

9.   The engine operationally Checked and a compressor wash done.

10.   An engine vibration analysis performed.

A turbine engine powered business or commuter aircraft might undergo a set of routine inspections to insure the engine remains in an airworthy condition. An example of a typical inspection schedule would be as follows:

Transit checks after each flight

These consist of checking the engine cowling for signs of leaks, the intake and low pressure compressor rotor blades for damage, the engine exhaust for damage and metal deposits, and checking the oil level and oil filter blockage indicator.

Check 1 at 25 hours of run time

This consists of the transit checks plus:

Checking the last stage of the low pressure turbine for cracking and damage by using a strong spotlight.

Checking the nose cone fairing for damage and security, the low pressure compressor lining and rotor blades, the low pressure compressor outlet guide vanes, and the intermediate compressor inlet guide vanes.

Check 2 at 150 hours or 3 months

This consists of the transit checks and check 1, plus:

Checking the two master chip detectors

Checking the fan bypass duct, exhaust cone and fairings for damage.

Audibly checking the igniter plugs.

Checking the low pressure and high pressure air supply ducting and joints for damage and leaks

Checking the fuel system for contamination.

Checking the freedom and security of mechanical controls and greasing those controls.

Check 3 at 600 hours or 12 months

This consists of the transit checks, and check 1 and check 2, plus:

Replacing the fuel filter element.

A functional test of the temperature control system. Large commercial aircraft, like Boeing 767's or **Airbus A340's,** undergo what is called a Continuous Airworthiness Inspection Program. This program involves daily inspections, known as layover inspections, and what are called letter- or alphabet checks. The letter checks are typically broken down into what are called "A", "B", "C" and "D" checks.

Each higher letter check incorporates all the previous inspection items, plus a more detailed look at the aircraft and its engines. By the time the "D" check takes place, the aircraft is ready to be completely stripped to its shell and rebuilt. What will happen to the engines at this point in time depends on how long it has been since they were overhauled. They may just be removed, inspected, and reinstalled as the aircraft goes back together.

Non Routine

During the operation of a gas-turbine engine, various events may occur which cause the engine to required an immediate special inspection to determine whether the engine has been damaged and what corrective actions must be taken. Among some of the events which may cause the engine to require special inspections are foreign-object, ingestion, bird ingestion, ice ingestion, over limit operation (temperature and rpm), excessive "G" loads, and any other event that could cause internal or external engine damage.

Non-routine inspections require the same techniques as those used for daily and periodic inspections. These techniques include unaided visual inspection, inspection with lights, use of magnifiers, application of fluorescent or dye penetrants, use of a borescope or videoscope, and use of radiography techniques. Usually the maintenance manual for the engine will specify which technique is the most effective for a particular inspection.

Unscheduled line maintenance is performed as the maintenance technician corrects discrepancies found during flight, walk-around inspections, scheduled inspections, while p6rforming Airworthiness Directives (AD's) etc. The engine system Chapters VI to XII of this text, include the line maintenance repairs and troubleshooting necessary to locate and correct common malfunctions.

Borescope, Fiberscope and Electronic Imaging

The borescope was used for many years as a device for examining the insides of cylinder bores on reciprocating engines and is now extensively used on turbine engines.

The borescope is a rigid instrument that may be compared with a small periscope. At one end is an eyepiece with one or more lenses attached to the light-carrying tube. At the end of the tube are a mirror, a lens, and strong light. The tube is inserted through engine borescope ports located in the engine case at points necessary to allow for examination of all critical areas inside the engine. The ports are normally closed (with removable plugs). One type of borescope system is shown in Figure.

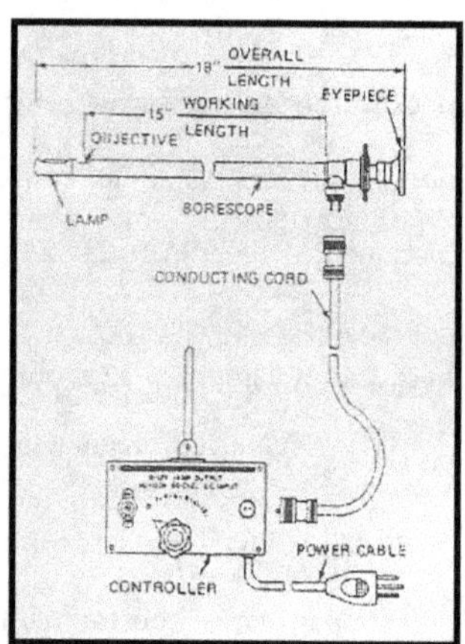

When borescope inspections are to be performed, the technician should identify the plugs as they are removed to be sure that they are reinstalled in the same ports. Upon reinstallation, the threads and pressure faces of the plugs should be lightly coated with an anti seize compound such as MIL-T-5544 or its equivalent.

A variation of the rigid borescope is the fiberscope, which is shown in Figure. The flexible fiberscope usually has a controllable bending section near the tip so that the observer can direct the scope after it has been inserted into an engine inspection port. This bending action allows the fiberscope to scan the area inside the engine once inside the port. Many times it is necessary to use the fiberscope to inspect around corners inside the engine when no inspection entry port is available to allow direct line of sight.

Correct identification of cracks, stress, and corrosion is critical during maintenance erections. Inspectors often find it difficult to differentiate between an actual defect an unclear image. A new imaging technique, electronic imaging, is able to produce true-color, magnified images that can be seen on a video monitor. One such system, the Video probe 2000 (manufactured by the Welch Allyn Company), is shown in Figure.

A video imaging system includes an inspection probe, a video processor, and a video monitor for displaying the image. The system uses a tiny charged-coupled device (CCD) sensor in the tip of the probe. The solid-state CCD sensor acts like 'a miniature TV camera to transmit the image electronically to a video monitor. First, light is transmitted to the inspection area, either by light-emitting diodes (LEDs) or by fiber-optic light guides, depending on the inspection probe selected. A fixed-focus lens in the tip of the probe gathers reflected light from the area and directs it to the surface of the CCD sensor. The signal then travels down the length of the probe through amplifiers. The video processor receives the signal, digitizes it, assembles it, and outputs it directly to a video monitor, video tape recorder, or computer enhancement equipment.

Video imaging lends itself to high-quality videotape and photographic documentation. The images can be viewed by several inspectors at different locations via multiple video monitors.

Foreign-Object Damage (FOD)

Foreign-object damage to a gas-turbine engine may consist of anything from small nicks and scratches to complete disablement or destruction of the engine. The flight crew of an aircraft may or may not be aware that FOD has occurred during a flight. If damage is substantial, however it will be indicated by vibration and by changes in the engine's normal operating parameters Damage to the compressors or turbines usually results in an increase in EGT, a decrease in engine pressure ratio (EPR), and a change in the rpm ratio between the core engine and the fan section ($N_2.N_1$ ratio).

When FOD has occurred, the inspections required depend on the nature of the foreign object or objects. If an external inspection indicates substantial damage to the fan section or to inlet guide vanes, the engine must be removed and overhauled. If the damage to the forward sections of the engine is slight, a borescope inspection of the interior of the engine may determine that it is unnecessary to remove the engine. Damage to vanes, fan blades, and compressor blades can be repaired if it does not exceed certain limits specified by the manufacturer. If the engine operates normally after repairs are made, it can be placed back in service.

Fan Blade Shingling

Fan blade shingling is the over-lapping of the mid-span shrouds of the fan blades. When the blades of a rotating fan encounter resistance which forces them sideways an appreciable distance, shingling will take palace Figure provides a simplified illustration of the situation which causes shingling.

Shingling can be caused by engine stall, bird strike, FOD, or engine overspeed in which case the fan must be inspected at both the upper and lower surfaces of the mid-span shrouds for chafing, scoring, and other damage adjacent to the interlock surfaces. All blades that are overlapped or show indications of overlapping must be removed and inspected according to the appropriate manual. No cracks are permitted in the fan blades. Blade tips are examined for curl/land the lightening holes are checked for cracks and deformation.

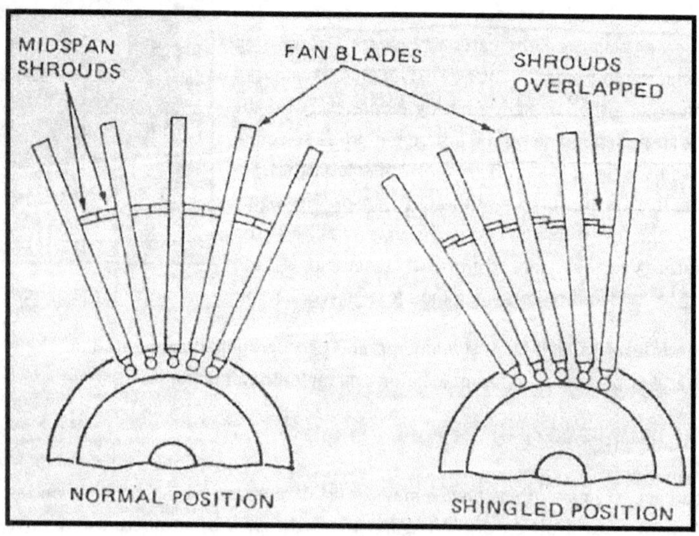

Additional inspections in the fan area include inspection of the abradable material for damage due to rubbing of fan blade tips and inspection of the fan-speed sensor head for damage due to blade contact.

Inspections for Overlimit Operation

Even though technicians and flight crews take every precaution possible to prevent over-limit operation of engines, such operation sometimes occurs. Often the cause is a malfunction of the engine fuel control or a malfunction in the engine. In any case, when over-limit operation does occur, it is necessary to perform certain inspections to determine what damage may have resulted.

At starting, the most critical parameter for the engine is EGT. The technician or crew member starting the engine must watch the EGT gage carefully As soon as lightoff occurs, there is a rapid rise in EGT; but if all systems are working prop the EGT should not exceed limits, If it does, the person starting the engine should immediately retard the start lever to reduce fuel flow to the combustion chamber.

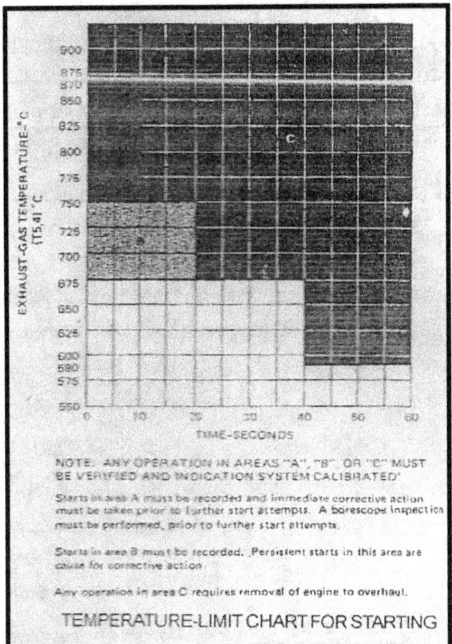

The technician who starts a gas-turbine engine should be familiar with the operating limitations. Figure shown is a temperature-limit chart for starting a large, high-bypass engine. Note that any temperature above 675°C is cause for special attention. Temperatures that fall in area A require special inspections, and temperature-time values that fall in area C are cause for engine overhaul.

After an engine has been started and the operation is stabilized at ground idle, higher temperatures can be permitted during taxiing and preparation for takeoff. The chart in Figure shows temperature-time limitations for operations ether than starting. The charts in Figures are applicable to one particular engine only and are not typical of all engine limitations.

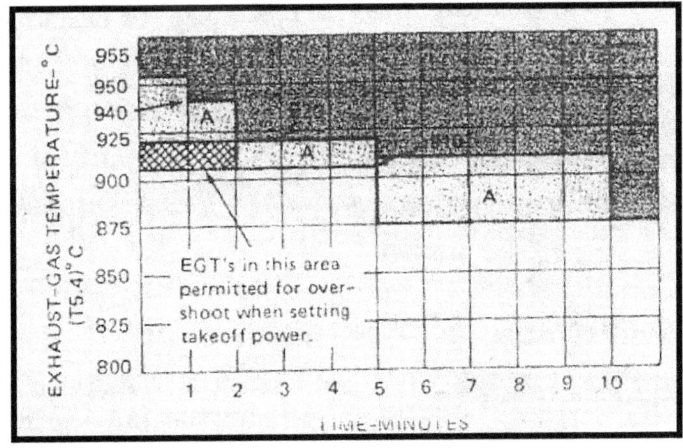

If a gas-turbine engine has been operated above the limits set for EGT but at a level not high enough to call for removal and overhaul, a borescope inspection is usually called for. An external visual inspection of the hot section of the engine should also be made. In this inspection, the hot section of the engine is checked for indications of burn-through or metal distortion due to excessive heat. If such indications are found, the section must be disassembled for further inspection and repair.

When borescope inspections are called for after operation at excessive temperatures, the following are inspected:

1. Combustion chamber and liner assembly to determine if cracks and burned areas exceed those permissible as specified by the manufacturer.

2. Fuel nozzles for excessive carbon buildup or plugged orifices.

3. First-stage high-pressure turbine (HPT) nozzle for cracks, burned areas, warping, and plugged cooling-air passages. Serviceable limits for defects specified in the Maintenance manual must be met.

4. Second-stage HPT nozzles for defects as listed.

5. HPT rotor for cracks, tears, nicks, dents, and metal loss. Cracks in the turbine blades are cause for removal and replacement. Dents and nicks within certain limits may be permitted in the second-stage blades, as specified by the manufacturer.

6. Turbine midframe liner for cracks, nicks, dents, burns, bulges, and gouges. Limitations for these defects are specified by the manufacturer. Bulges associated with heat discoloration are cause for rejection.

7. First-stage low-pressure turbine (LPT) nozzle for cracks, nicks, dents, bums, etc., as for other turbine sections.

8. LPT stator assembly as above.

9. LPT rotor assembly as above. No cracks are permitted in any turbine blades. Limited dents and nicks are allowed.

Overspeed inspections

Overspeed inspection for a typical high-bypass fan engine is primarily concerned with rotating assemblies. One manufacturer specifies the following inspections if the fan section has been operated at speeds from 116 to 120 percent rpm:

1. Check the fan rotor for freedom of rotation.

2. Check the first-stage fan shroud for excessive rub.

3.   Inspect the low-pressure compressor with a borescope.

4.   Inspect the inlet and the exhaust nozzles for particles.

5.   Inspect all four stages of the LPT with a borescope for blade and vane damage. Inspect the fourth-stage blades through the exhaust nozzle.

If the fan speed has exceeded 120 percent, the fan rotor, fan midshaft, and LPT rotor must be removed, disassembled, and inspected in accordance with instructions.

If the core-engine rotor (high-pressure compressor and high-pressure turbine) has been operated at speeds from 107 to 108.5 percent, the following inspections are specified:

1.   Inspect the exhaust nozzle for particles.

2.   Inspect the core compressor with a borescope for blade and vane damage.

3.   With a bores cope, inspect the HPT for blade carnage

If the core engine rotor has been operated above 108.5 percent, the engine must be removed, disassembled, and inspected according to instructions.

1.   Turbine Wheel

Turbine blade blend repair limits which are typical of either shop or flight line maintenance. According to this manufacturer, cracks are never acceptable (which is the case with most manufacturers).

Of particular concern during visual inspections are stress rupture cracks on turbine blade leading or trailing edges. Stress rupture cracks are perceptible as minute hairline cracks at right angles to the blade length. This condition and rippling of the trailing edge is indication of a serious over temperature, and a special in-shop manufacturer's inspection will probably be required.

To maintain turbine wheel balance, a single turbine blade replacement is generally accomplished by installing a new blade of equal moment-weight. If the blade's moment-weight cannot be matched, the damaged blade and one at 180° out are replaced with blades of equal weight; or, the damaged blade and the blades 120° from it are replaced with three blades of equal moment weight in the same manner as was mentioned for compressor blades.

Turbine blades are rarely replaced on the wing today, however, shop procedures usually allow for entire turbine reblading, after which the rotor is checked on a special balancing device.

Code letters indicating the moment-weight in inch-ounces or inch-grams, are marked on the fir-tree section of the blade. The turbine blades are checked for creep and twist.

a.   Creep and Untwist

Creep is a term used to describe the permanent elongation which occurs to rotating part Cree is most pronounced in turbine blades because of the heat loads and centrifugal loads imposed during operation Each time a turbine blade is heated, rotated, then stopped (referred to as an en me cycle), it remains slightly longer than it was before The additional length may be only millionths of an inch under normal circumstances or, after an engine over temperature or over speed condition, very much longer.

Nevertheless, if the blade remains in service long enough, chances are that it will eventually make contact with its shroud ring and begin to wear away When this occurs, an audible rubbing can be heard on engine coast-down. Clearance checks are then taken and appropriate maintenance action is determined.

Creep can progressed in three stages; Primary, secondary, and tertiary. The primary and tertiary stages occur relatively quickly. Primary creep occurs during the engine's first run, tertiary during operating overloads. But the secondary creep stage occurs quite slowly. It is within the secondary creep region that the engine manufacturer bases the turbine's service life.

Accelerated (tertiary) creep during the engine's service life can be attributed to the following:

A.   Hot starts/over temperatures;

Extended operation at high power (high EGT and centrifugal loading);

Erosion of the blades from ingestion of sand or other foreign objects.

Untwist occurs in both turbine blades and turbine vanes from gas loads upon their surfaces, loss of correct pitch affects efficient of the turbine system, and engine performance deterioration results. The check for untwist is generally only possible after engine tear down when pats can be measured in special shop fixtures.

B.   Shop Maintenance (Heavy)

Whenever the engine cannot be repaired in the airplane, it is removed for shop maintenance or for test cell operation and troubleshooting. The Federal Aviation Administration (FAA) requires that this level of maintenance be accomplished only at a manufacturer's facility or at a certified repair station which has the necessary tooling, technical data, and trained personnel.

The FAA divides heavy maintenance for most engines into the two categories of limited and unlimited.

1.   Limited Heavy Maintenance

Many privately owned repair stations and factory- operated repair stations are in the limited category. They normally are authorized to perform any maintenance up to removal and replacement of the entire hot section. They can perform some cold end repairs, but cannot rebuild the compressor.

2.   Unlimited Heavy Maintenance

Privately owned and factory operated overhaul facilities fall into this category. They can remove and replace any part, perform limited remanufacture of parts, and zero out the engine time.

### 3.	Power Plant Removal

Power plants are fully removed from the aircraft by one of two methods. The method involves lowering the engine from its mounting location using a hydraulically operated installation stand which looks similar to a large scissors jack. This method is generally used when working with large engines. The other method, more common to general aviation, requires a sling and hoist arrangement to lower the engine into its transpiration dolly.

### 4.	Shop Maintenance

Once ill the shop, the engine to be repaired is usually installed on a maintenance stand. Some stands are on casters. Many stands are designed to keep the engine perfectly horizontal: The turbine, as well as many of the large components, will be placed in its own roll-around maintenance stand.

At some point engine disassembly is usually completed vertically, such as for removal of the compressor. On some engines, the entire disassembly is accomplished vertically so the weight will assist in component alignment. One standard maintenance practice during disassembly is to cover all openings as they are exposed, using plugs, caps, and other suitable material to prevent contamination and to maintain the utmost in shop cleanliness and safety procedures.

Another general rule that applies during any maintenance is never to reuse lock wire, lock washers, tab locks, cotter pins, gaskets, packings, or rubber O-rings and to reuse locknuts and other fasteners only with-in the limits prescribed in the manufacturer's instructions.

We see an expanded view of an entire General Electric CJ-610 engine, as it would 3.-7::\ear disassembled during heavy maintenance.

Hot Section Inspections

The hot section inspection is called on the basis of completed hours or cycles whichever comes first.

A hot section inspection is needed to determine the integrity of the components in the hot section of the engine. The hot section of the engine consists of the combustion section (burner cans or liners), turbine inlet guide vanes, turbine wheels, and related parts. Hot section inspection intervals vary widely depending on TBO (time between overhauls) and engine service experience. Generally, hot section inspections are performed on either a "time-in-service" basis or an "on-condition" basis. When engine parameters start to deteriorate, this is also evidence that a hot section inspection is needed. Whenever engine operational conditions dictate that an inspection is needed, it should be performed regardless of the time in service. Such conditions include engine overspeed, sudden stoppage, lighting strike, loss of oil, or unusual noises. An operational performance check (record of engine parameters) should be made before and after the hot section inspection to determine the improvement in engine performance.

In order to perform a hot section inspection, the engine will have to be somewhat disassembled. However, before this can be done, parts of the ignition system and fuel system will need to be disassembled and removed from the engine. Proper references and special tools needed should always be obtained before the disassembly procedure begins. The condition of the engine's components as well as certain clearances and dimensions, which are outlined in the disassembly and inspection sections of the maintenance manual, may need to be recorded during the disassembly phase of the inspection. During the inspection phase, the condition of integral engine components such as the turbine and combustion chamber sections will be closely examined and their airworthiness evaluated. At this time, faulty parts are rejected and replaced or sent to a certified repair station for reworking. During engine reassembly, it is critical that the manufacturer's instructions be carefully followed. All clearances and torque values must be observed. After completion of the reassembly phase, all paperwork should be completed, and a complete inventory of all tools and materials should be made.

In this text it would be impossible to cover all the hot section inspections for the many different types of turbine engines used in aviation. However, an attempt will be made to cover one example of a typical hot section inspection.

During hot section inspection the following parts are to be replaced

1. Combustion Chambers

2. Turbine blades

3. Fuel Nozzles

4. Stage - 1 NGV's

## Chapter - 7

ENGINE STOR AGE

INTRODUCTION

Under normal operating conditions the interior parts of an engine are protected against corrosion by the continuous application of lubricating oil, and operating temperatures are sufficient to dispel any moisture which may tend to form; after shutdown the residual film of oil gives protection for a short period. When not in regular service, however, parts which have been exposed to the products of combustion, and internal parts in contact with acidic oil, are prone to corrosion. If engines are expected to be out of use for an extended period they should be ground run periodically or some form of anti-corrosive treatment applied internally and externally to prevent deterioration.

The type of protection applied to an engine depends on how long it is expected to be out of service, if it is installed in an aircraft, and if it can be turned.

This Leaflet gives guidance on the procedures which are 'generally adopted to prevent corrosion in engines but, if different procedures are specified in the approved Maintenance Manual for the particular engine, the manufacturer's recommendations should be followed.

INSTALLED TURBINE ENGINES

Installed turbine engines which are to be out of use for a period of up to seven days require no protection apart from fitting covers or blanks to the intake, exhaust and any other apertures, to prevent the ingress of dust, rain, snow, etc. A turbine engine should not normally be ground run solely for the purpose of preservation, since the number of temperature cycles to which it is subjected is a factor in limiting its life. For storage periods in excess of seven days additional precautions may be necessary to prevent corrosion.

Short-term Storage - The following procedure will normally be satisfactory for a storage period of up to one month.

Fuel System - The fuel lines and components mounted on the engine must be protected from the corrosion which may result from water held in suspension in the fuel. The methods used to inhibit the fuel system depend on the condition of the engine and whether it is installed in an aircraft or not, and are fully described in the appropriate Maintenance Manual. On completion of inhibiting, the fuel cocks must be turned off.

Lubrication Systems. Some manufacturers recommend that all lubrication systems (engine oil, gearbox oil, starter oil, etc.) of an installed engine should be drained, and any filters' removed and cleaned, while others recommend that the systems should be filled to the normal level with clean system oil or storage oil. The method recommended for a particular engine should be ascertained from the appropriate Maintenance Manual.

External Treatment. Exterior surfaces should be cleaned as necessary to detect corrosion, then dried with compressed air. Any corrosion should be removed, affected area's retreated, and any damaged paintwork made good in accordance with the manufacturer's instructions. Desiccant or vapour phase inhibitor should be inserted in the intake and exhaust, and all apertures should be fitted with approved covers or blanks.

Long-term Storage. For the protection of turbine engines which may be in storage for up to six months, the short-term preservation should be applied a d, in addition, the following actions taken: -

    (i)   Grease all control rods and fittings.

    (ii)   Blank-off all vents and apertures on the engine, wrap greaseproof paper round all rubber parts which may be affected by the preservative and spray a thin coat of external protective over the whole engine forward of the exhaust unit.

At the end of each successive six months storage period an installed engine should bE re-preserved for a further period of storage. Alternatively, the engine may be removed from the aircraft and preserved in a moisture vapour proof envelope.

UNINSTALLED ENGINES (PISTON AND TURBINE)

Engines which have been removed from aircraft for storage, or uninstalled engines which are being returned for repair or overhaul, should be protected internally, and sealed in moisture vapour proof (MVP) envelopes. This is the most satisfactory method of preventing corrosion, and is essential when engines are to be transported overseas.

A piston engine should be drained of all oil, the cylinders inhibited, drives and inside of crankcase sprayed with cylinder protective, and all openings sealed:

A turbine engine should be drained of all oil, fuel system inhibited, oil system treated as recommended by the manufacturer, and blanks fitted to all openings.

Particular care should be taken to ensure that no fluids are leaking from the engine, and that all sharp projections, such as locking wire ends, are suitably padded to prevent damage to the envelope.

The MVP envelope should be inspected to ensure that it is undamaged, and placed in position in the engine stand or around the engine, as appropriate. The engine should then be placed in the stand, care being taken not to damage the envelope at the points where the material is trapped between the engine attachment points and the stand bearers.

Vapour phase inhibitor or desiccant should be installed in the quantities and at the positions specified in the relevant Maintenance Manual, and a humidity indicator should be located in an easily visible position in the envelope. The envelope should then be sealed (usually by adhesive) as soon as possible after exposure of the desiccant or vapour phase inhibitor.

The humidity indicator should be inspected after 24 hours to ensure that the humidity is within limits (i.e. the indicator has not turned pink). An unsafe reading would necessitate replacement of the desiccant and an examination of the MVP envelope for damage or deterioration. After a period of three years storage in an envelope the engine should be inspected for corrosion and re-reserved.

INSPECTION

Engines in storage should be inspected periodically to ensure that no deterioration has taken place.

Engines which are not preserved in a sealed envelope should be inspected at approximately two-weekly intervals. Any corrosion patches should be removed and the protective treatment re-applied, but if external corrosion is extensive a thorough inspection may be necessary.

Envelopes on sealed engines should be inspected at approximately monthly intervals to ensure that humidity within the envelope is satisfactory. If the indicator has turned pink the envelope should be unsealed, the desiccant renewed and the envelope resealed.

Equipment. The spraying equipment should be of a type approved by the engine manufacturer, and should be operated in accordance with the instructions issued by the manufacturer of the equipment. For inhibiting cylinders a special nozzle is required, and this should be checked immediately before use to ensure that the spray holes are unblocked. Correct operation of the spray gun may be checked by spraying a dummy cylinder and inspecting the resultant distribution of fluid.

Materials. Only the types of storage and inhibiting oil recommended by the manufacturer should be used for preserving an engine. American manufacturers generally recommend oils and compounds to American specifications, and British manufacturers generally recommend storage oil to DEF 2181, wax-thickened cylinder protective to DTD 791, turbine fuel system inhibiting oil to D. Eng. R.D. 2490, and external air drying varnish approved under a DTD 900 specification. Only approved alternatives should be used, and any instructions supplied by the manufacturer in respect of thinning or mixing of oils should be carefully followed.

Blanks. Approved blanks or seals should be used whenever possible. These are normally supplied with a new or reconditioned engine, and should be retained for future use. Pipe connections are usually sealed by means of a screw-type plug or cap such as AGS 3802 to 3807, and plain holes are sealed with plugs such as AGS 2108; these items are usually coloured for visual identification. Large openings such as air intakes are usually fitted with a specially designed blanking plate secured by the normal attachment nuts, and the contact areas should be smeared with grease before fitting, to prevent the entry of moisture. Adhesive tape may be used to secure waxed paper where no other protection is provided, but should never be used as a means of blanking off by itself, since it may promote corrosion and clog small holes or threads.

REMOVAL FROM STORAGE

For an engine which was not installed in an aircraft during storage the installation procedure described in the appropriate Maintenance Manual should be carried out, followed by a thorough ground run and check of associated systems. For an engine which was installed in an aircraft during storage the following actions should be taken: -

    (i)   Remove all masking, blanks and desiccant.

    (ii)   Clean the engine as necessary, e.g. remove excess external protective and surplus grease from controls.

    (iii)   Ensure fire extinguisher spray pipe holes are clear.

    (iv)   Replace any components which were removed for individual storage, de-inhibiting as necessary.

    (v)   Drain out all storage oil, clean oil filters and refill with normal operating oil.

    (vi)   Piston engines; remove sparking plug blanks and turn engine slowly to drain excess oil from the cylinders, then fit plugs and connect leads. Turbine engines; prime the fuel system in accordance with the manufacturer's requirements.

    (vii)   Prime the engine lubricating oil system.

    (viii) Start the engine and carry out a check of the engine and associated systems.

RECORDS

Appropriate entries must be made in the engine log book giving particulars of inhibiting procedures or periodic ground running. Such entries must be signed and dated by an appropriately licensed engineer or Approved Inspector.

## Chapter - 8

POWERPLANT INSTALLATIONS AND REMOVAL PROCEDURE

INSTALLATION OF GAS-TURBINE ENGINES

Introduction

The installation and mounting of gas-turbine engines in aircraft involve a number of factors, all of which affect the performance or reliability M the engine. Among these factors are the designs of the air-inlet system, exhaust system, cooling arrangement, and mounting structure: These elements are largely the responsibility of the airframe manufacturer; however, the engine manufacturer is also involved because he must supply the engine data which enable the airframe manufacturer to design a suitable installation. Exhaust systems, including thrust reversers, noise suppressors, and conventional nozzles, are the concern of both the engine manufacturer and the airframe manufacturer. The design, however, must be such that it adapts itself to the needs of the aircraft in which the engine is installed.

Air Inlets

When it is realized that a modern turbojet engine may devour as much as 350 to 550 tons of air per hour, the importance of efficient air-inlet-duct design immediately becomes apparent. Because of structural considerations and the placement of cockpits and equipment, the efficiency of the air-inlet duct has suffered

It times; however, the development of high-pressure-ratio engines, with their sensitivity to changes in airflow, has made efficient inlet-duct design a prime objective. It is the purpose of this section to explore some of the characteristics of air-inlet ducts and to point out the advantages and disadvantages of various types.

Importance of Structural Design

Figure shows the effect of air-inlet-duct efficiency on the thrust of a turbojet engine at various speeds. It is immediately evident that inefficient duct design takes a heavy toll, especially as airspeed increases. It will be noted that, for a given duct efficiency, the losses at 600 mph are about three times as great as those at 300 mph. In addition to duct losses, serious performance penalties may also result from the external effects of the inlets, either because of added form drag or because of prematurely induced compressibility effects. These conditions also become more pronounced as speed increases.

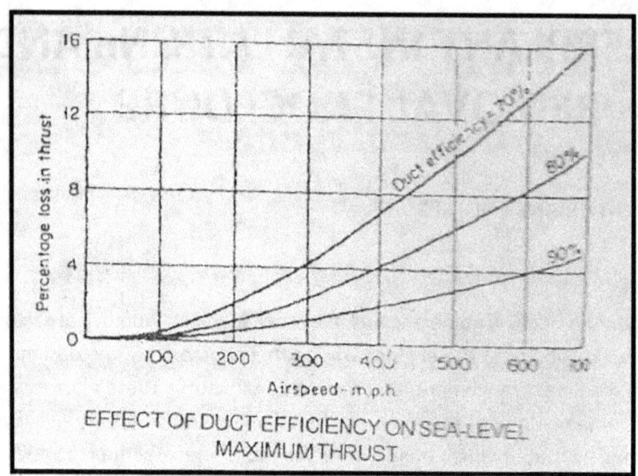

EFFECT OF DUCT EFFICIENCY ON SEA-LEVEL MAXIMUM THRUST

is indicated that the aerodynamic requirements for the ideal air-induction system are (1) that it delivers air to the engine with no reduction in available energy, (2) that it does not increase the drag of the body on which it is placed, (3) that it does not compromise the high-speed characteristics of the airplane, and (4) that it provides adequate air induction in ail reasonable flight attitudes. To these requirements might be added, in the case of axial-flow installations, the necessity of maintaining a fairly uniform pressure distribution over the compressor face.

Unfortunately, these criteria must be weighed against structural and design considerations in arriving at the details of any given installation. For this reason, we have almost as many variations in induction systems as we have airplanes, each one of which meets the requirements for an ideal system to a greater or lesser degree, depending upon the extent of compromise necessary on overcoming particular design problems.

Types of air inlets

In the pasta wide variety of air inlets have been employed or considered for jet engines installed in aircraft. Five types which were considered possible for subsonic fighter aircraft are shown in the drawings of Figure. Of the five types shown, the nose inlet and wing root inlet have been found to be most suitable.

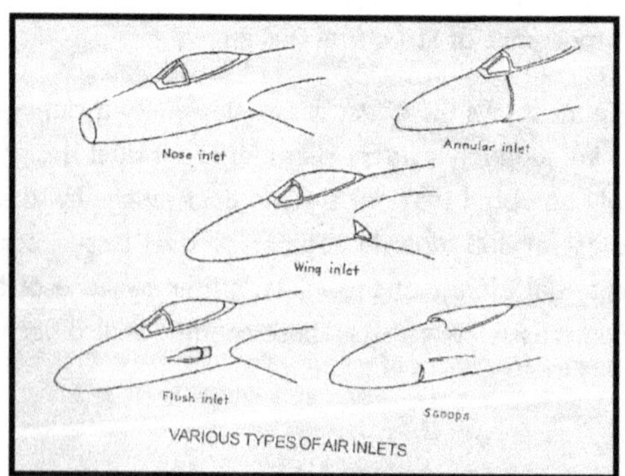

VARIOUS TYPES OF AIR INLETS

76

For subsonic aircraft such as the F-100 the nose inlet was found to be quite satisfactory. The inlet adds no aerodynamic drag to the aircraft, and it permits a reasonably direct flow of air to the compressor of the engine. The disadvantages of this type of duct lie in the length of the duct and the difficulty in providing adequate strength. The nose-inlet duct also offers problems when it comes to locating the cockpit and various items of required equipment.

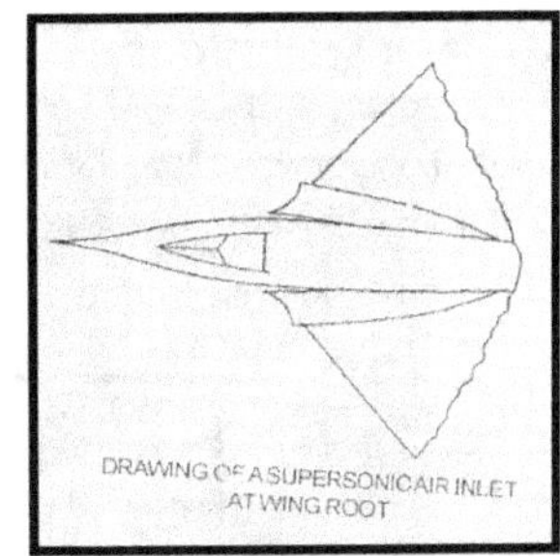

Modern fighter aircraft such as the F-102, F-104, F-105, and F-106 utilize wing root air inlets. These inlets are especially designed to meet the requirements of supersonic flight and the effects of shock waves which form at the inlets. The inlets are set out 4ightiv with an open gap between the inlet and the fuselage to eliminate the effects of the boundary layer. A general idea of the arrangement of such air inlets is shown in the diagram of Figure.

ENGINE PODS SUSPENDED FROM THE WING

The simplest and most efficient air inlet is provided by the pod engine mounting used for modern jet airliners. These pods, whether suspended from the wing as shown in Figure, provide a no-drag airflow directly into the compressor of the engine. As the engine pods are mounted on struts to allow an air gap between the pod and the wing or fuselage, there is a minimum of aerodynamic interference with the smooth flow of air over the structure adjacent to the engine pod.

ENGINE PODS AT SIDE OF FUSELAGE

Typical Installations

A typical powerplant installation in an early fighter-type airplane is shown in Figure. In this installation, a circular nose inlet branches into two ducts just inside the nose. The ducts join just ahead of the engine to form an annular section that attaches to the compressor inlet.

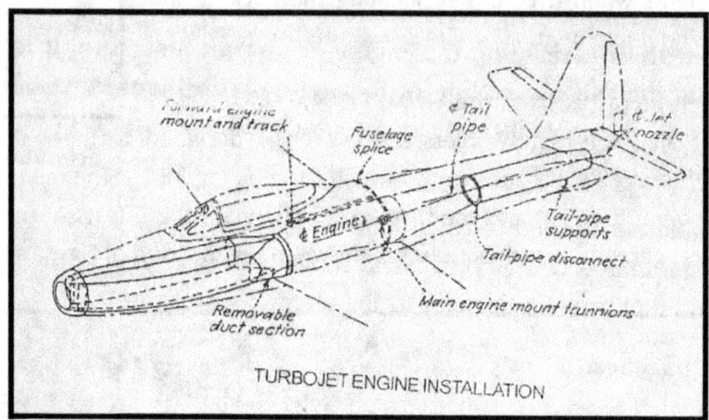

TURBOJET ENGINE INSTALLATION

The engine extends from the compressor inlet back to the quick-disconnect clamp that fastens the engine exhaust cone to the tailpipe. The tail-pipe terminates in a jet nozzle located immediately forward of the aft end of the fuselage, as shown in Figure, to form the primary of a cooling-air ejector.

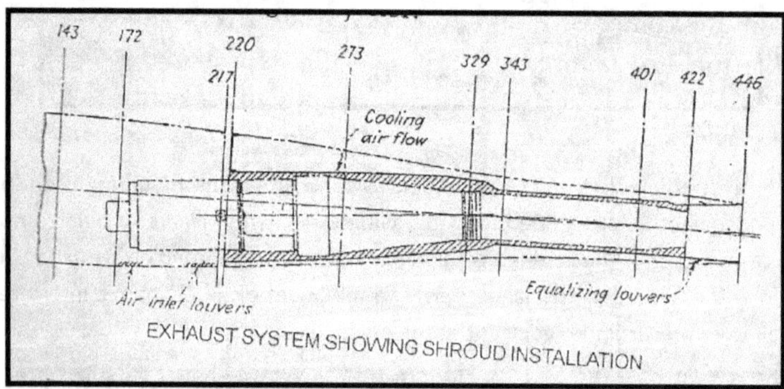

EXHAUST SYSTEM SHOWING SHROUD INSTALLATION

Cooling air enters the airplane through the air-inlet louvers and flows rearward between a stainless-steel shroud and the hot surfaces of the powerplant. No other insulation is used for the engine, exhaust cone; tailpipe, or shroud. This method of protecting the fuselage structure from the high exhaust-gas temperatures was chosen mainly for two reasons: (1) a comparison of shroud weights with the weight of insulating blankets indicated that a shroud would be lighter; and (2) an ejector was used rather than a ram scoop in order to assure sufficient airflow for cooling and burner compartment ventilation during ground operation.

Investigations have shown that it is not a good practice to cool the exhaust system too much. The hot gases produced by the engine contain considerable energy in the form of heat. If the gases are cooled before they leave the jet nozzle, they lose energy-and the engine loses thrust. It is therefore desirable to retain as much heat as possible in the jet stream without damaging the engine or aircraft structure. It is common practice to insulate the exhaust section of the engine with blankets constructed of metal screen and foil. These blankets retain the heat in the gases and prevent it from escaping to the aircraft structure.

Aircraft structures exposed to high temperature are commonly constructed of stainless steel or titanium. It is thus possible to retain a large degree of the heat in the exhaust stream and still prevent damage to the aircraft structure.

Firewall or bulkhead

As systems were developed and put into operation, the requirement for an external bulkhead to separate the compressor and turbine sections was established. This bulkhead is commonly called a firewall. Its purpose is to prevent fuel from entering the region around the combustion, turbine, and exhaust sections in the event that fuel leakage occurs in the compressor and accessory sections of the engine as a result of battle damage or fuel-line failure.

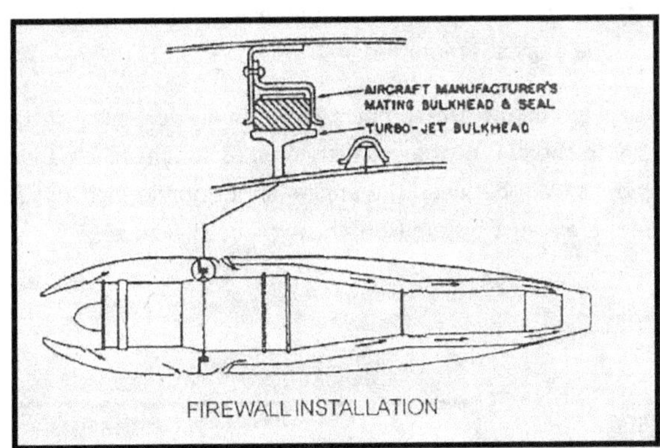

The firewall bulkhead on the engine and the mating bulkhead provided by the aircraft manufacturer are shown schematically in Figure. The addition of the bulkhead across the nacelle at this point requires that the cooling air-inlet louvers for the aft section of the nacelle be located on the rear side of the bulkhead. Louvers are also generally provided ahead of the bulk-head, through which is, discharged the ventilating air from the compressor section.

Problems of design

Experience has shown that the problems of exhaust-system design require careful study and analysis in order to obtain the safest and most efficient configuration.

Because of the pressure existing within the tailpipe during operating, an axial force, acting rearward, tends to force the pipe to the rear, away from the engine. The engine is designed to withstand this axial force, the load being carried through the exhaust cone and into the engine structure. The rearward force acts against the engine thrust; hence it subtracts from the net thrust. For this reason, the tailpipe must be designed to create a minimum of 'reverse thrust.

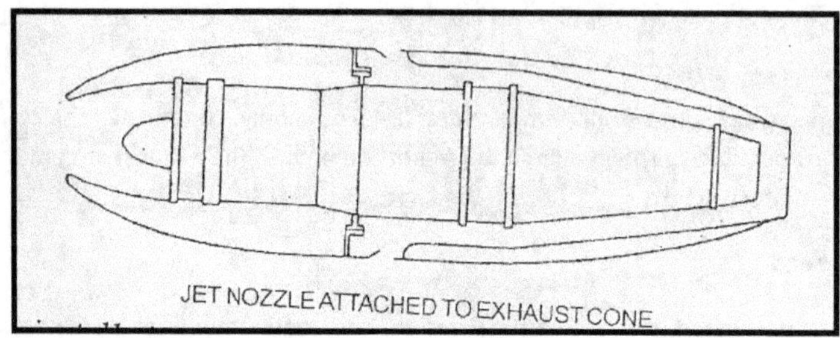

For engine installations with thrust reversers, a more complex arrangement is required for both structural design and control system arrangement. The thrust reverser is mounted in such a manner that the reverse loads are carried through the engine to the engine mounts and thence to the wing or other aircraft structure to which the engine is mounted.

The exhaust-system design problems are comparatively simple for those installations where the jet nozzle is fastened directly to the end of the exhaust cone, as shown schematically in Figure; or where only a short length of pipe is used between the engine an jet nozzle, because for these cases it is not necessary to support the rift; exhaust system from the airplane structure.

Tailpipe mounting

Another problem becomes evident when the tailpipe mounting provisions are considered. The pipe may be too long and heavy to be cantilevered from the engine exhaust cone and then must be supported from the airplane structure. This would be fairly simple except for fuselage deflections that occur because of control surface and landing loads. It is therefore necessary in some cases to make a flexible attachment between the pipe and the cone. Figure shows a joint used on a

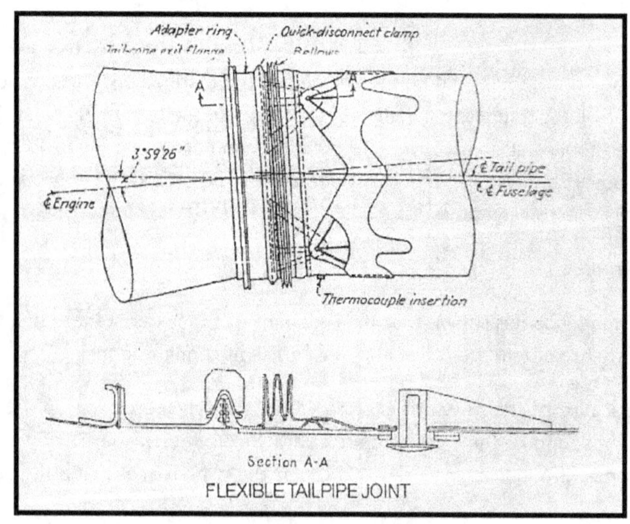

typical airplane. The normal tension loads in the pipe are carried through four pins and load-distributing straps, with the seal being made by means of a bellows. Clearance is allowed in the pin bearings so that the pipe can bend sufficiently to allow for the maximum fuselage deflections. The flexible joint also permits expansion and contraction without distortion.

TYPICAL ENGINE INSTALLATIONS

The actual procedure for installing gas-turbine engines in aircraft must in all cases follow the instructions provided by the manufacturer in the Aircraft Maintenance and Overhaul Manual. As noted in the preceding portion of this chapter, there are many different types of installations, depending upon the purpose for which the aircraft is designed and the type of engine installed.

Military aircraft engine installations

An example of how a turbojet engine is installed in a military fighter aircraft is shown in Figure. As shown in the photograph, the airplane fuselage can be separated and the engine removed or installed quickly and easily. The photograph illustrates an F-100 fighter plane with a J57 engine. The engine is equipped with an afterburner, which is a requirement for the majority of combat aircraft.

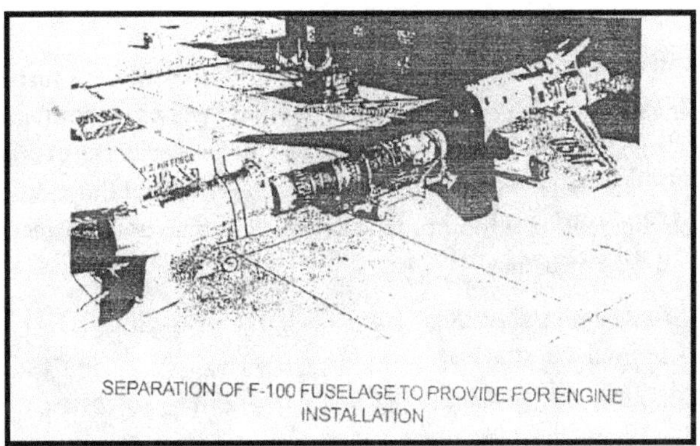

The design of military jet engine installations is such that an engine can be replaced in a very short time. With some aircraft, an expert crew can remove and replace an engine in less than fifteen minutes. This is made possible by the fact that the airplane fuselage can be separated and that all system connections (fuel, pneumatic, electrical, mechanical, etc.) are of the quick-disconnect type. The engine change equipment, such as engine dollies, hoists, and tail dollies, is designed to make the operation quick and easy.

Figure illustrates the ease with which a crew can perform maintenance on an engine when the rear section of the fuselage is removed. The crew in the picture is working on the after-burner section of a 157 engine.

Mounting for a turboprop engine

An example of, the type of engine mounting employed for a turboprop engine is shown in Figure. These drawings show the tubular steel engine mount together with enlarged views of the attachment fittings for the Rolls-Royce Dart engine installed on a Fairchild "Friendship" F-27 airliner. It will be observed that the mount is similar to a typical mount for a reciprocating engine and consists of six steel tubes welded to seven fittings to form a W-shaped structure.

Three fittings, welded at the three forward points, are attached to pedestal-type mounting brackets installed on the engine intermediate compressor casing. Each engine mount fitting is secured to a mounting bracket by a longi-tudinal hollow tension bolt that passes through a seating washer and the mounting bracket and is then threaded into the tapped engine mount fitting. The hollow tension bolt is locked in place by a concentric left-hand threaded locking bolt that is also threaded into the engine mount fitting. A lock ring or tabwasher, installed under the head of the locking bolt, engages the hollow-tension bolt and maintains bolt torque.

The four aft fittings support the powerplant on the forward face of the firewall. Each fitting is attached to a firewall mounting bracket with transverse AN bolts in shear. The engine mount tube structure is utilized to support the engine control linkage, fire detector cable, electrical harnesses, fuel heater, and various hose and heat shield assemblies.

Pod mounting for a turbofan engine

The installation of a Pratt & Whitney JT3D turbofan engine on a Boeing 720 airliner is illustrated in Fig. This illustration shows the left side of a typical engine installation and reveals a number of the details associated with such an installation.

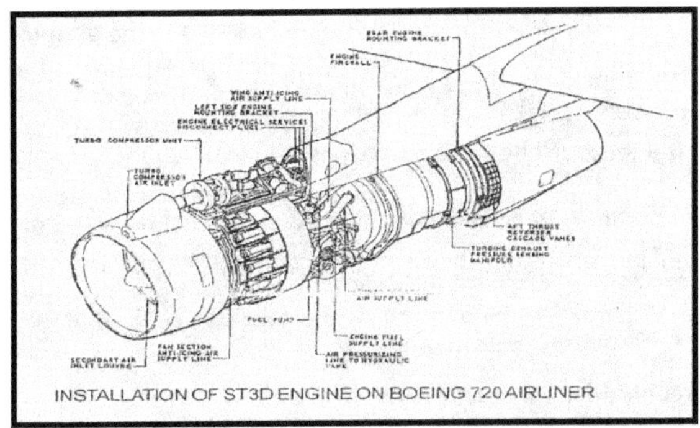

INSTALLATION OF ST3D ENGINE ON BOEING 720 AIRLINER

The engine is secured to the strut at three points. The forward mounting position has an arrangement of links bolted to the forward flange of the diffuser case and terminating in two cone bolts which are rigidly secured into the strut fitting. The aft engine mount is comprised of two links bolted to the top of a double flange around the exhaust section. These terminate in a single cone bolt secured into a strut-mounted bracket arranged to compensate for changes in engine length due to thermal expansion. The mount fittings and their locations are shown in the drawings of Figure.

ENGINE REMOVALAND INSTALLATION

New or reconditioned turbine engines are normally supplied as an Engine Change Unit 4ECU), the Unit including the basic engine and the equipment which is common engines on the

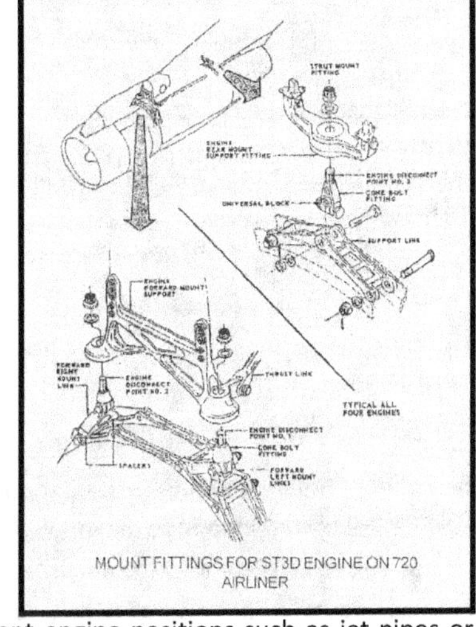

MOUNT FITTINGS FOR ST3D ENGINE ON 720 AIRLINER

particular aircraft. Items which are 'handed' to suit the different engine positions such as jet pipes or engine mountings, and other items which are not fitted to all engines such as thrust reversers or hydraulic pumps, must be added to the ECU to make it into a complete power plant. The transfer of these components from the old power plant to the new ECU, when approved, must be recorded in the appropriate log books. Similarly, when new components are fitted to the ECU the entries must be made in the ECU log book. Engine installation details vary considerably between aircraft and it is not within the scope of this Leaflet to present detailed information on each. The following paragraphs may be taken to be typical of current practices and should be read in conjunction with the appropriate Maintenance Manual Where the term 'engine' has been used in the text this may equally be taken to imply 'ECU' or 'Power Plant' as appropriate.

Engine Removal

Some aircraft may have to be jacked up and trestled when removing an engine. This is not often practicable with large multi-engine aircraft and other means are sought to ensure that the engine is at the same attitude when suspended from the sling as when installed in the airframe. Manufacturers may specify that the aircraft should be placed in a certain attitude (e.g. 1°30' nose down) for engine removal and it is also sometimes necessary to add ballast weights to one side of an engine which is not symmetrical. These requirements will be stated in the aircraft Maintenance Manual.

NOTE: If the engine is not being replaced immediately it is sometimes recommended that ballast weights be fitted to the engine mountings to prevent airframe distortion.

Preparation

The aircraft should be prepared for engine removal as follows:

a.   Ensure that the landing gear ground lock pins are correctly fitted.

b.   Turn off the fuel supply to the engine being removed.

c.   Disconnect all electrical power to the engine being removed.

d.   Fit blanks to the engine air intakes and jet pipe.

e.   Position lifting gear. f. Prepare and position a suitable engine stand ready to receive the engine after removal.

Safety Precautions

The maximum safe working load of the lifting gear must exceed the weight to be lifted and the sling must be the correct one for the work being carried out (i.e. a sling designed to lift the bare engine must not be used to lift a complete power plant). The sling should be inspected for frayed wires, bent or worn shackles and any other damage likely to affect its serviceability. An aircraft undergoing engine change will rise or fall as the weight changes. All personnel working near the aircraft should be made aware of this and the proximity of trestles and stands checked to avoid structural damage.

Removal

Sufficient cowlings and panels should be removed to gain access to the engine disconnect points, engine slinging points and engine mountings. All system connections should then be broken, the open ends of pipes and electrical connectors being fitted with blanks to prevent the ingress of dust and dirt. Controls and cables should he temporarily secured to prevent damage; slave pulleys are sometimes provided in the nacelle to receive detached control cables. Locking wire, split pins, washers, nuts or bolts which have been removed should be carefully collected and taken away from the engine location. It is good practice to keep an inventory of tools used by the operators to check that none are left where they could cause damage.

NOTE: The electrical charge held by high energy ignition units can be lethal. After removing the LT input lead at least one minute should elapse before touching the HT output lead or igniter plug.

The engine sling should be fitted and the weight of the engine taken on the lifting gear, attaching ballast weights as necessary to level the engine. After a final check to ensure that all the connections are broken, the mountings should be disconnected and engine lowered into the prepared stand.

Any components required for the replacement engine should be removed, together with the engine sling, and the engine prepared for storage or transit.

ENGINE INSTALLATION

Preparation

With the engine in its stand an inspection should be made for any damage which could have occurred in transit. Components being installed on the engine should be inspected for serviceability, special note being taken of any lubrication or testing required by the manufacturer. The engine bay area should be thoroughly cleaned and inspected for damage and all engine connections inspected to ensure that no damage was sustained during engine removal. Engine mountings should be lubricated in accordance with the manufacturer's instructions.

The safety precautions appertaining to slings and lifting gear, for removal, apply equally to installation and must be checked before commencing lifting operations.

Installation

The engine sling should be fitted and the engine lifted from the stand into the engine bay, connecting the mountings and torque loading the attachment bolts to the values quoted in the Maintenance Manual. When the engine is secured in position the sling should be removed and all systems reconnected, removing the blanks immediately before connecting each item. The high energy ignition unit should not be connected until the engine is ready for ground running.

All bolts should be tightened to the recommended torque values and new gaskets, washers and locking devices fitted as necessary. Particular attention must be paid to the setting up of engine controls, the procedure for which will be found in the appropriate Maintenance Manual. A duplicate inspection of these controls is required.

Priming

The engine oil tank should be filled with the approved oil, and the oil and fuel systems primed. This is to ensure that all inhibiting oil is removed, that all pipelines are full and that the engine will not rotate in dry bearings when started.

(a)   Oil System - Using the approved type of oil and a priming rig, the micro-pumps, gearbox and oil filters should be primed with the quantity of oil specified in the Maintenance Manual followed by replacement and locking of all unions, plugs and filler caps.

(b)   Fuel System - As the aircraft booster pump is used for this operation, electrical power to the engine should be restored and the low pressure fuel cock opened. The booster pump should then be switched on and the appropriate bleed valves opened until clean bubble-free fuel is discharged. The bleed fuel should be collected in a suitable container and the bleed re-locked after use.

NOTE: It is common airline practice to cross-feed the fuel supply when bleeding. This ensures complete purging of the low pressure supply.

Preparation for Starting

To prove the engine installation before starting, the following procedure should be carried out:

(a)   Check that the engine installation is clean and free from extraneous material, remove the intake and jet pipe covers and place drip trays under the engine.

(b)   Open the low pressure fuel cock (LP cock) and the high pressure fuel cock (HP cock) and carry out a motoring run with the ignition system switched off. Check that oil and fuel pressures are indicated and that there are no leaks from the engine or adjacent pipelines.

(c)   Allow the engine to drain and carry out a second motoring run with the HP cock closed.

(d)   Allow the engine to drain once more then connect the leads to the high energy ignition units and refit all cowlings and panels. The engine should then be started, all systems tested and adjustments carried out as necessary.

## Chapter -9

GAS- TURBINE OPERATION
INSPECTION, TROUBLESHOOTING
MAINTENANCE AND OVERHAUL

Preparation for Starting

To prove the engine installation before starting, the following procedure should be carried out:

(a)   Check that the engine installation is clean and free from extraneous material, remove the intake and jet pipe covers and place drip trays under the engine.

(b)   Open the low pressure fuel cock (LP cock) and the high pressure fuel cock (HP cock) and carry out a motoring run with the ignition system switched off. Check that oil and fuel pressures are indicated and that there are no leaks from the engine or adjacent pipelines.

(c)   Allow the engine to drain and carry out a second motoring run with the HP cock closed.

(d)   Allow the engine to drain once more then connect the leads to the high energy ignition units and refit all cowlings and panels.

The engine should then be started, all systems tested and adjustments carried out as necessary.

GROUND RUNNING

The life of a turbine engine is affected both by the number of temperature cycles to which it is subjected and by operation in a dusty or polluted atmosphere. Engine running on the ground should therefore be confined to the following occasions:

(a)   After engine installation.

(b)   To confirm a reported engine fault.

(c)   To check an aircraft system.

(d)   To prove an adjustment or component change.

(e)   To prove the engine installation after a period of idleness.

STARTING AND OPERATION

Starting Gas-Turbine Engines

A gas-turbine engine should be started only when all conditions required for the safety of the engine and nearby property and personnel are met. The engine pod or nacelle should be checked for loss material, tools, or other items which could be ingested by the engine and cause foreign-object damage (FOD) to vanes, blades, and other interior parts. The best type of surface on which to operate gas-turbine engines mounted in aircraft is smooth concrete that is free of all items or material with could heat and high-velocity gases will not cause damage to other aircraft, ground service equipment, or vehicles and will not cause injury to personnel.

Figure shows the temperatures and velocities of the exhaust stream exiting a JT9D engine at both idle and takeoff speeds. Even at idle speed, the temperatures and gas velocities are hazardous. At takeoff power, the engine produces dangerous temperatures and high gas velocities more than 200 ft to the rear of the jet nozzle. Observe in the drawings that a 25-ft radius at the front of the engine is considered a danger zone because of the velocity of the air flowing to the engine inlet. This area should be carefully avoided by maintenance personnel and should be kept clean to prevent foreign objects from being drawn into the engine's intake. There have been .7.r.ariv reported instances of inlet ingestion of personnel working around operating engines, demonstrating the serious consequences of violating the hazard zones delineated in most maintenance manuals. The inlet airflow velocity is low at the hazard

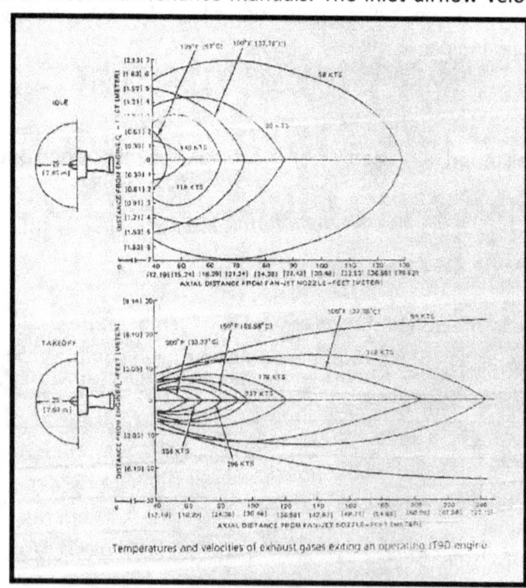

Temperatures and velocities of exhaust gases exiting an operating JT9D engine.

boundary, but it rapidly increases to the point of no recovery as the inlet is approached. Personnel should not approach the engine's inlet by moving forward alongside the engine's cowling. The airflow that moves forward along the sides of the nacelle will increase greatly near the engine's inlet.

The principles involved in the starting of a gas-turbine engine are relatively simple. It is merely necessary to rotate the engine at a speed sufficient to provide adequate air volume and velocity for starting, provide high-intensity ignition in the combustion chamber, and introduce fuel through the fuel nozzles in an amount that will not produce excessive heat but will provide sustained combustion and further acceleration of the engine.

Small gas-turbine engines are often equipped with starter-generators. These units are electric motors which apply starting torque for the engine during starting; then, when the engine rpm attains a self-sustaining level, the starter motor becomes a generator to supply the aircraft electrical system. The starters draw a high level of current when starting and therefore heat up very rapidly, they are thus subject to damage by overheating if their operation is not limited to the time intervals specified

Large gas-turbine engines are generally equipped with air-turbine starters which receive their air supply from a ground service unit (air cart), another engine, or an auxiliary power unit (APU) installed in the aircraft.

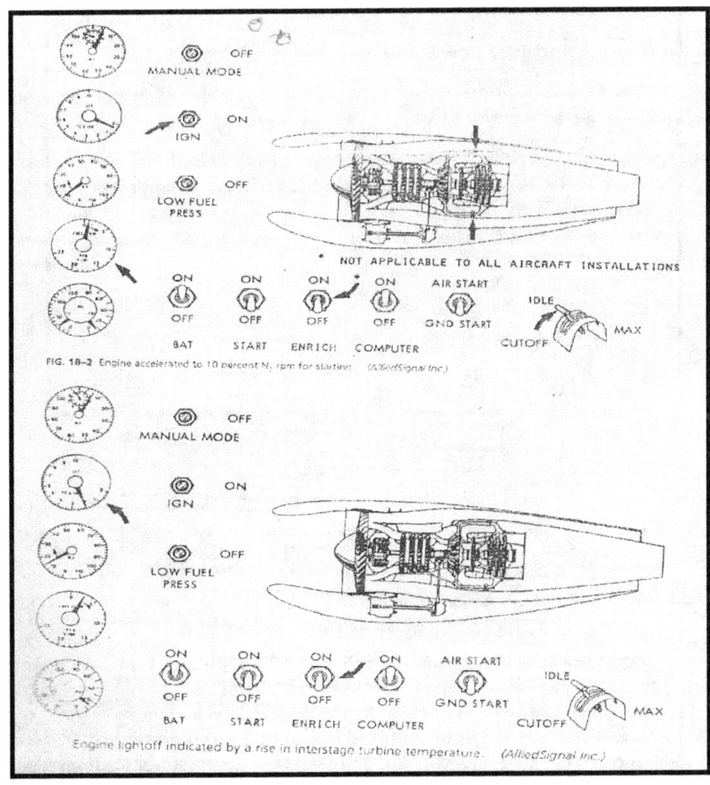

FIG. 18-2 Engine accelerated to 10 percent N, rpm for starting. (AlliedSignal Inc.)

Engine lightoff indicated by a rise in interstage turbine temperature. (AlliedSignal Inc.)

The starting of gas-turbine engines can be manual or automatic, depending on the particular engine installation. The majority of starts for today's engines are accomplished automatically.

The manual start of a gas-turbine engine usually includes the following steps :

1.   Conned the starting power unit to the aircraft, if used.

2.   Provide for electrical power on the aircraft.

3.   Turn on the fuel and fuel boost pumps.

4.   Turn on or arm the ignition circuit. The ignition system is often energized by a. switch connected to the start lever or power lever.

5.   Check the starting system to ensure that there is enough starting energy for a complete start cycle. Engage the starter and accelerate the engine until the high-pressure compressor (N_2) or gas generator speed is between 10 and 25 percent rpm. This speed will vary among different engines. In this example, the speed is 10 percent rpm.

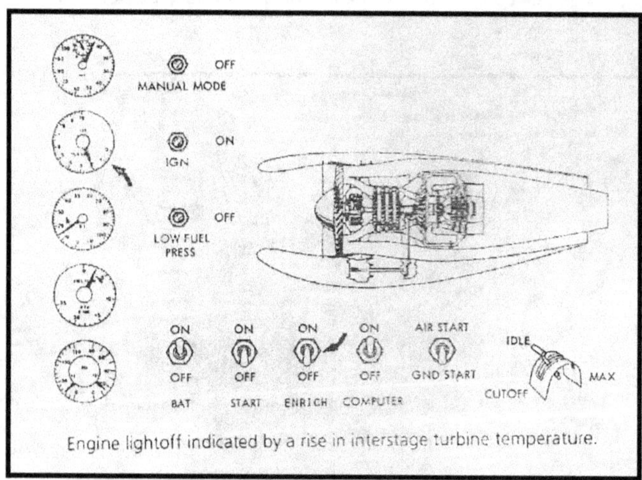

Engine lightoff indicated by a rise in interstage turbine temperature.

6. Advance the power lev0 or start lever depending on the type of engine being started. On some engines, the power lever is set in the IDLE position for starting and the start lever controls fuel flow to the engine. Note the fuel flow gage and check for a stabilized fuel flow. Now watch the interstage turbine temperature (ITT) or exhaust-gas temperature (EGT) indicator until a rapid temperature rise is indicated, as shown in figure. This signifies a lightoff (fuel burning in the combustion section). When a start lever is used, it is generally moved to the START or IDLE position in one movement. If the power lever is used for starting, do not advance the power lever further until the temperature stabilizes. The temperature must not be permitted to exceed the maximum allowable specified for the engine. If the temperature climbs too high, the fuel flow must be shut off to prevent a hot start condition. A hot start

can burn and destroy the engine's internal parts. If the EGT rises above the specified amount, the fuel flow should be stopped and the engine turned to draw cool air through it. If no rise in EGT is detected, then the condition is a no-lightoff condition. This is generally caused by a faulty ignition system or an incorrect fuel control start fuel-air mixture.

7. If lightoff has taken place and the engine's EGT has stabilized, the EGT will begin to decrease as engine rpm increases. Advance the power lever until the engine rpm reaches idle speed, but do not permit the temperature to exceed the level specified.

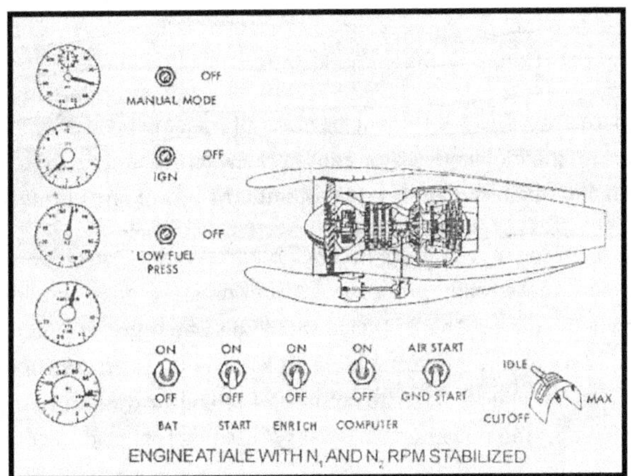

8. The starter should be released after the operator is certain that the engine has reached a self-sustaining speed, as shown in figure. The starter cutout can be automatic or manual depending on the aircraft being started. It is important that the starter is cut out at the proper percent rpm to prevent overspeeding and damaging of the starter. To start a gas-turbine engine with an automatic system, the operator follow the procedure established for the particular aircraft and engine. An automatic start may be accomplished as follows, with some variations depending on the aircraft and engine combination.

1. See that an air supply is available from a ground service unit, an onboard APU, or an operating engine on the aircraft.

2. Turn on the aircraft electric power.

3. Turn on the fuel and fuel pump.

4. Place the .power lever in the IDLE position.

5. Rotate the engine selector switch to the position for starting the engine selected.

6.   Press the starter switch. It is normally equipped with a holding coll so that it will remain closed until the engine is rotating at a self-sustaining speed. At this point, current to the holding coil is cut off and the switch opens.

The engine accelerates when it reaches a level of N_2 rpm Somewhat above 10 percent of maximum, fuel will be supplied under pressure to the fuel nozzles. The ignition is turned on automatically at the time that the starter switch is depressed or shortly thereafter, but it is always on before fuel flow is supplied. Soon after fuel flow begins (within 20s,) fuel is ignited and the engine accelerates at an increased rate. Fuel flow is controlled automatically by the fuel control unit and is not permitted to exceed the amount required for the correct rate of acceleration. Normally an engine will not have a hot start or a hung start unless there is a defect in the starting system or fuel control.

When the engine has accelerated to 35 percent or more of maximum rated rpm, a centrifugal switch cuts off electric power to the holding coil of the start switch, and the switch opens. The engine accelerates to idle speed and remains in that condition until the power lever is moved by the operator.

It is important that the operator of a gas-turbine engine observe the instruments pertaining to engine operation while performing the starting procedure. A fuel pressure gage should indicate correct fuel pressure before the engine is started. The EGT gage should be watched closely at lightoff to ensure that the temperature has not exceeded the maximum value allowed Oil pressure should register on the oil pressure gage shortly after the engine starts to rotate. If the engine does not accelerate properly or if the EGT exceeds the safe limit, the power-control lever should be retarded to cut off the fuel and stop the engine

If, during the starting of a gas-turbine engine, the EGT exceeds the prescribed safe limit, the engine is said to have had a hot start. When this occurs, the engine fuel flow should be shut 'down immediately Continue to rotate the engine with the starter to provide some engine cooling Hot starts are usually caused by an excess of fuel entering the combustion chamber (malfunctioning fuel control) or insufficient starting power (low battery or air pressure). In an engine with an automatic fuel control unit, hot starts normally do not occur unless the unit is malfunctioning.

If the engine fails to accelerate properly or does not reach the idle rpm position the starting attempt is called a false start, or a start. The automatic systems used on modem engines prevent this type of problem from occurring, as long as the fuel control unit is functioning properly and sufficient energy is supplied to the starter. Either a hung start or a hot start can be caused by an attempt to ignite the fuel before the engine has been accelerated sufficiently by the starter. Another type of problem is not having sufficient energy to turn the starter to complete the starting process.

Typical Airline Start Procedure

Technicians and flight-crew members employed by an airline are given specific instructions and checklists which they must follow before and during an engine start. The procedures to be followed in the starting of a JT8D engine on a 727 aircraft are as follows:

1. Perform a walk-around inspection according to the appropriate checklist.

2. Perform a control cabin prestart check as set forth in the checklist.

3. Start the APU. (An APU control panel is shown in figure)

    a. Turn essential bus-selector switch to APU.

    b. Switch ac meter to APU.

    c. Aim automatic fire shutdown switch.

    d. Reset APU fire detection system.

    e. Turn master switch to ON. f. After 10 s, turn master switch to START and release to ON position.

    g. Observe for light off and EGT max.

    h. Check APU frequency and voltage (400 Hz, 115 V c) and close APU generator breaker to power the aircraft.

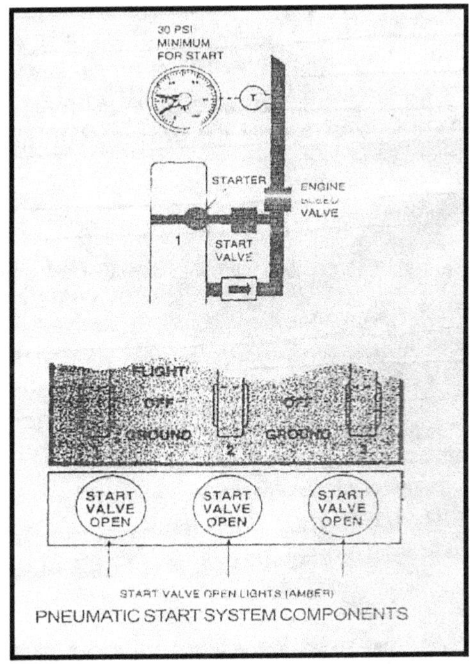

PNEUMATIC START SYSTEM COMPONENTS

4. Obtain "All Clear" signal.

5. Check pneumatic pressure (35 psi minimum).

6.   Set engine start switch to GROUND START, shown in figure. This will allow the start valve to be opened and pneumatic pressure to turn the air-turbine starter which turns the N_2 compressor. The starter will stay engaged until it is released, later in the starting process.

7.   When N_2 rpm is 20 percent, place the start lever in IDLE position, as shown in figure.

8.   Check for N_1 rotation and low oil pressure light out.

9.   Observe fuel flow rates and starting EGT. EGT should not be more than 662°F when outside air temperature (OAT) is less than 59°F or not more than 788°F when OAT is more than 59°F. or not more than 788°F when OAT is more than 59°F.

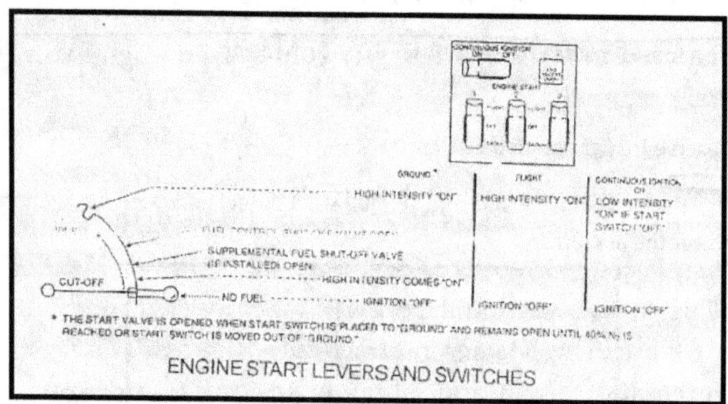

ENGINE START LEVERS AND SWITCHES

10.   Release start switch at 40 percent N speed.

11.   Check the following engine parameters after idle speed stabilizes:

EGT	300 to 420°C when air bleed or power extraction is used.
N2 rpm	54 to 59.4
Oil pressure	40 to 55 psi (44 to 46 psi desired)
Oil temperature	40 to 60°C (120°C Maximum)
Fuel flow	Approximately 1000 lb/h

Cold Weather Engine Operation

In the start-up of, a cold-soaked turbine engine, some basic steps are essential, such as ensuring that there is an -adequate air supply, dual ignition, and the correct fuel mixture. The start valve can stick closed because of ice; this can generally be corrected by applying heat to the valve area.

To start the engine, motorize it to as high a speed as possible, ensuring an N_2 speed of at least 10 percent, and select ENRICH. If the engine fails to light up within 90 s, select FUEL OFF. Ice can accumulate around the fan area, which will prevent fan rotation during starting. This is one of the reasons that the fan rotation (N) must be checked during engine starting. The temperature indication of a lightoff will likely be much lower in extremely low temperatures. Under an extreme cold soak, such as sub zero (°F) temperature conditions, oil pressure to, a full-scale reading The Filter Blocked" warning light may illuminate, and the indicated oil quantity may fall to zero. As oil temperature rises, with, the engine at idle, with the engine at idle, oil pressure will fall back, the "Filter Blocked" light will go off, and oil quantity will return to normal. The correct procedures to be used in starting a severely cold-soaked engine can be found in the engine's operating manual.

2. Hot Section

For hot section a variety of high strength to weight materials have been developed, often referred to as super alloys. These alloys have a maximum temperature limit of 2000°F when unclooled and 26000F when cooled internally. Super alloys were developed for use in high tem-perature areas where oxidation resistance is needed and where high thermal, tensile, and vibratory stresses arc present.

Super alloys are complex mixtures of many critical metals, namely: Nickel chromium, cobalt titanium, tungsten, carbon, and other metallic elements.

General agreement on precise mixtures by manufacturers of turbine parts is still the subject of much debate. One reason for this is that the strength properties these metals ultimately have depends on the mixture. However, the stronger the metal, the more difficult and expensive it is to form and machine into the intricate shapes necessary for turbine engine parts.

Also, exotic materials and processes mean expensive initial and replacement costs. It is not uncommon to see a single, two to three inch long turbine blade, that costs $300 or a large blade that costs $3,000. The technician's decision to change these parts soon becomes a significant cost to the customer.

Many manufacturing processes are utilized in the production of hot section parts. Forging, casting, and plating are the traditional methods. Newer procedures include powder metallurgy, single-crystal casting, and plasma spraying.

Powder metallurgy is a forging process in which powdered super metal is hot-pressed into a solid state. This results in a very high density material of high temperature strength. These metals have long slender crystals with axial grain boundaries but few traverse grain boundaries and are very creep resistant.

Another process which produces even stronger materials is single crystal casting in which only one grain of material forms in the mold rather than millions of grains bound together as is the case with traditional casting. With only one grain and no grain boundaries, corrosion due to expansion is all but eliminated.

Ceramic and aluminium alloy thermal barrier coating of super alloy parts and some titanium parts are also processes which give high surface strength and resistance to corrosion. These coatings are generally referred to as plasma sprays and, when applied under high heat, melt into the surface of the base metal. This coating is said to give the best protection against the scaling type corrosion or erosion which occurs at high gas temperatures. Scaling is a condition caused by sodium (salt) in the air and sulfur in the fuel reacting chemically with the base metals.

The alloy itself, in conjunction with the process involved in forming it, usually determines its strength and workability. In some cases, the combustion liner is constructed of material as thin as 0.040 inch and must be easily weldable. For this reason, nickel-base alloys are generally used today rather than the stainless steel alloys of a few years ago. Nickel-base alloys contain little or no iron are non-corrosive and can be worked in thin weldable sheets.

Combustion liners sometimes experience surface erosion which supports carbon buildup on the inner surfaces. Today one might see a whitish coating applied to some liners, called magnesium zirconate. It is a material which helps combat carbon deposits. A property of this coating is that it wears away during operation taking carbon with it and thereby maintaining cleanliness on the surfaces. During heavy repair the coating is replaced.

Combustor cases and turbine cases are often constructed from nickel-base alloys, often referred to as Inconel (a trade name).

Turbine blades and vanes are either forged by newer powder metallurgy technique or by traditional methods or investment cast from nickel-base alloys. They are also cast methods mentioned previously. These materials have very high temperature strength under centrifugal loads and are highly corrosion resistant.

Turbine disks are almost exclusively constructed from cobalt base alloys for high thermal strength under extreme mechanical and thermal loads. Temperatures as high as 2500^0F are experienced in the first stage vane area of large engines. Cobalt, representing the current state-of-the-art in metallurgical development for turbine engines, is used to combat these high temperatures.

Because of their ability to withstand higher heat than metal, and also their high strength to weight ratio, ceramic turbine components have been experimented with since early German development; but, because of its low mechanical strength under vibrating loads, only limited use has been made of ceramic material. Some non-flight engines are currently fitted with ceramic blades and disks, but flight engines to date are limited to ceramic coating of some stationary parts, such as combustion liners and turbine nozzle vanes. Ceramic coating can usually be recognized as a greenish glazed type finish.

As mentioned previously the gas turbine engine could have unlimited power if there were no temperature limits imposed by material strength. The idea is that less and less cooling air would be needed as material heat strength increases. The engine then, in turn could be scaled down and the power to weight ratio would increase significant-ly. But material strength today. as it was in Whittle's time, still remains the most limiting factor in the power output of the turbine engine.

GAS-TURBINE ENGINE OVERHAUL

In the past, most engines had specified numbers of hours they could operate before they needed to be overhauled. This period became known as the time between overhauls (TBO).

The length of time between overhauls varies widely with different types of engines. When a new type of engine enters service, its TBO is fairly short, but as condition monitoring, the engine's service record, and inspections prove the engine to be reliable, the TBO is generally extended. Many engines have proven to be so reliable that they are overhauled only when they need major maintenance. This concept is a form of "on condition" maintenance or overhaul.

Because the TBO is actually determined by the life of one or two assemblies within the engine, during overhaul it is generally found that the other assemblies are mechanically sound and fit to continue in service for a much longer period. Therefore, with the introduction of modular engines and the improved inspection and monitoring - techniques available, the TBO method of limiting the engine's life on-wing has been replaced by the "on condition" method. Basically this means that a life is not declared for the total engine but only for certain parts of the engine. On reaching their life limits, these parts are replaced and the engine continues in service, with the remainder of the engine being overhauled "on condition." Modularly constructed engines are particularly suited to this method, because the module containing a life-limited part can be replaced by a similar module and the engine returned to service with minimum delay. The module is then returned to the manufacturer or the overhaul shop and is disassembled for life-limited part replacement, repair, or complete overhaul, as required.

The overhaul of gas-turbine engines is accomplished by the manufacturer or at approved overhaul stations. The process is similar in many ways to the overhaul of piston engines; however, there are processes required which are not necessary for piston engines. In addition, an overhaul facility for gas-turbine engines requires many special tools and some equipment specially designed for work on particular types and models of engines.

The overhaul of a gas-turbine engine includes a complete disassembly and inspection. Non repairable parts are discarded, and those salvageable through rework` or recycling are sent to an appropriate facility. Repairable parts are processed and then given a rigid inspection and/or test to ensure that they are serviceable.

The average certificated aviation maintenance technician is not required to perform turbine-engine overhaul but can perform various field repairs as specified by the maintenance manual. The principal consideration for the technician is to be sure to have the correct manuals, 'bulletins, and other instructions when servicing and repairing a gas-turbine engine.

OVERHAUL PROCEDURES

Disassembly

The engine can be disassembled in the vertical or horizontal position. When it is disassembled in the vertical position, the engine is moulted, usually front end downward, on a floor fixture. To enable it to be disassembled horizontally, the engine is mounted in a special turnover stand.

The engine is disassembled into main subassemblies or modules, which are fitted in separate stands and sent to other areas where they are further disassembled into individual parts. The individual parts are taken to a cleaning area in preparation for inspection.

Cleaning

The cleaning agents used during overhaul range from organic solvents to acids, and other chemical cleaners, and extend to electrolytic cleaning solutions. Organic solvents include kerosene for washing, trichloroethane for degreasing, and paint-stripping solutions which can generally be used on the majority of components for carbon and paint removal. The more restricted and sometimes rigidly controlled acids and other chemical cleaners are used for removing corrosion, heat scale, and carbon from certain components. To achieve the highest degree of cleanness needed to perform the detailed inspection that is considered necessary for certain major rotating parts, such as turbine disks, electrolytic cleaning solutions are often used.

Inspections

After the components have been cleaned, they are visually and, when necessary, dimensionally inspected to establish their general condition and then further inspected for cracks. Inspection for cracks includes magnification, magnetic, or penetrant inspection techniques, used either alone or consecutively, depending on the components being inspected and the degree of inspection necessary.

Dimensional inspection consists of measuring specific components to ensure that they are within the limits and tolerances given in the Table of Limits. Some of the components are measured at each overhaul because only a small amount of wear or distortion is permissible. Other components are measured only when the condition found during visual inspection requires dimensional verification. The tolerances laid down for overhaul, supported by service experience, are often wider than those used during original manufacture.

Repair

To ensure that costs-are maintained at the lowest possible level, a wide variety of techniques is used to repair engine parts to make them suitable for further service. Welding, fitting of interference sleeves or liners, machining, and electroplating are some of the techniques employed during repair.

Some repair methods, such as welding, may affect the properties of the materials, and, to restore the materials to a satisfactory condition, it may be necessary to heat-treat the parts to remove the stresses, reduce the hardness of the weld area, or restore the strength of the material in the heat-affected area. Heat-treatment techniques are also used for removing distortions after welding. The parts are heated to a temperature sufficient to remove the stresses, and, during the heat-treatment process, fixtures are often used to ensure that the parts maintain their correct configurations.

Electroplating methods are also widely used for repair purposes. These methods range from chromium plating, which can be used to provide a very hard surface, to application of thin coatings of copper or silver plating, which can be applied to such areas as bearing locations on a shaft to restore a fitting diameter that is only slightly worn.

Many repairs are effected by machining diameters and/or faces to undersize dimensions or boring to oversize dimensions and then fitting shims, liners, or metal spray coatings of wear-resistant material. The affected surfaces are then restored to their original dimensions by machining or grinding. .

The inspection of parts after they have been repaired consists mainly of penetrant, or magnetic inspection. However, further inspection may be required for parts that have been extensively repaired; this inspection may involve pressure testing or x-ray inspection of welded areas.

Balancing

Because of the high rotational speeds, any unbalance in the main rotating assembly of a gas-turbine engine is capable of producing vibrations and stresses hi increase as the square of the rotational speed. Therefore, very accurate balancing of the rotating assembly is necessary.

The two main methods of measuring and correcting unbalance are single-plane (static) balancing and two-plane (dynamic) balancing. Single-plane balancing is used when the unbalance is in one plane only; that is, the unbalance goes centrally through component at 90° to the axis. The single-plane method is appropriate for components as individual compressors and turbine disks. For compressor assemblies and turbine-rotor assemblies possessing appreciable axial length, unbalance may be present at many positions along the axis; therefore, two-plane balancing may be required.

Assembling

The engine can be built up in the vertical or horizontal position using a ram, or stand. Assembly of the engine subassemblies, or modules, is done in separate areas, thus minimizing the build time on the rams.

Engine Testing

The testing of a new or overhauled gas-turbine engine to ensure correct performance is accomplished on an instrumented test stand. Procedures for testing are developed and published by the engine manufacturer, and these procedures must be followed precisely to ensure that correct information is obtained regarding the performance of the engine.

The operation of an engine on a test stand is usually accomplished with a bellmouth air inlet. The purpose of this type of inlet is to eliminate any loss of air pressure at the compressor inlet. The reason for loss of pressure with a straight inlet and the effect of the bellmouth inlet are illustrated in figure. Since a large volume of air is drawn into the engine, a rapid increase in air velocity must take place as the air nears the inlet. Moreover, to supply the demand, air must flow from areas outside the area directly in front of the engine. Much of the airflow will have to change direction almost 90° as it comes from the sides of the inlet and enters the compressor. With a straight inlet duct, this directional change results in a pressure drop. However, the bell mouth duct guides the air in such a way that there is essentially no pressure drop at the compressor inlet. If the bellmouth duct is protected by a screen, a certain amount of pressure drop will occur and must be taken into consideration when the performance of the engine is measured.

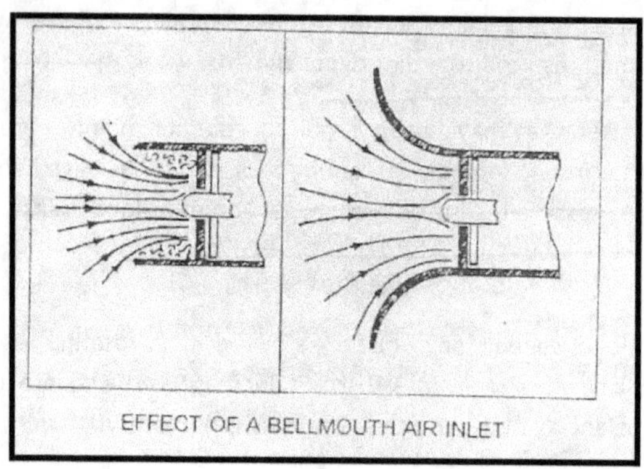

In testing of a gas-turbine engine, it is common practice to measure certain essential parameters in order to evaluate the engine performance correctly. Among these parameters are the following:

1. Ambient air temperature (T_{amb})

2. Ambient air pressure (P_{amb})

3. Exhaust total pressure (P_c)

4. Low-pressure compressor rpm (N_1)

5. High-pressure compressor rpm (N_2)

6. Exhaust-gas temperature (EGT)

7. Fuel flow in pounds per hour (pph) (W_j)

8. Thrust (F_n)

9. Low-pressure compressor outlet pressure (P_{s3})

10. High-pressure compressor outlet pressure (p^{s4})

These parameters are usually adequate to determine engine performance, but others may be recorded if desired or necessary.

When an engine is assembled as a complete powerplant for a quick engine change (QEC), it is necessary to consider the equipment installed on the engine, because it may affect some of the performance measurements. Oil flow and temperature will be changed as a result of the engine oil cooler and the engine pump. Likewise, fuel flow and pressures will be affected by the engine-driven fuel pump.

Because standard performance of an engine occurs only under standard conditions, air pressure and temperature must be corrected to standard conditions. This is accomplished by means of correction factors designated by the Greek letters delta (8) and theta (8). Delta is used to correct for pressure and theta provides the correction for temperature.

The values for delta and theta may be found on an appropriate chart or they calculated as follows:

$$d = \frac{P}{P_0} = \frac{P}{29.92}$$

$$q = \frac{T}{T_0} = \frac{t(^0F) + 460}{519}$$

where; P = observed barometric pressure (in HG abs)

P_0 = standard-day barometric pressure

T = temperature, ^0R ($^0F + 460$)

T_O = standard-day temperature, 519°R

If Kelvin degrees are used to indicate absolute temperature, then the Celsius or centigrade scale is used. Adding 273 converts centigrade to Kelvin. Standard-day temperature in degrees Kelvin is 288.

To apply d and q to the correction of measurements, the following methods are employed:

$$N_2 \text{ (corrected)} = \frac{N_2 \text{ (observed)}}{\sqrt{\theta_?}}$$

$$EGT\ ^0R \text{ (Corrected)} = \frac{EGT \text{ (observe)} + 460}{\sqrt{\theta_{t2}}}$$

$$W_f \text{ (corrected)} = \frac{W_f \text{ (Observed)}}{\sqrt{\theta_{t2}}}$$

$$F_n \text{ (corrected)} = \frac{Fn \text{ (observed)} + \text{Inst. corrected} + \text{Cell } corrected}{\delta_{t2}}$$

The values observed and calculated as shown for N_2 rpm, EGT, and fuel flow are recorded and plotted on a chart. Note that the pressure ratios for the low-pressure compressor and high-pressure compressor are recorded. These pressure ratios are indicated as P_{s3}/P_{amb} for the low-pressure compressor and P_{s4}/P_{amb} for the high-pressure compressor. The net thrust (Fn) of an engine can be determined directly from the thrust meter in the cell and can also be found from the EPR (engine pressure ratio) and an EPR conversion table for the engine.

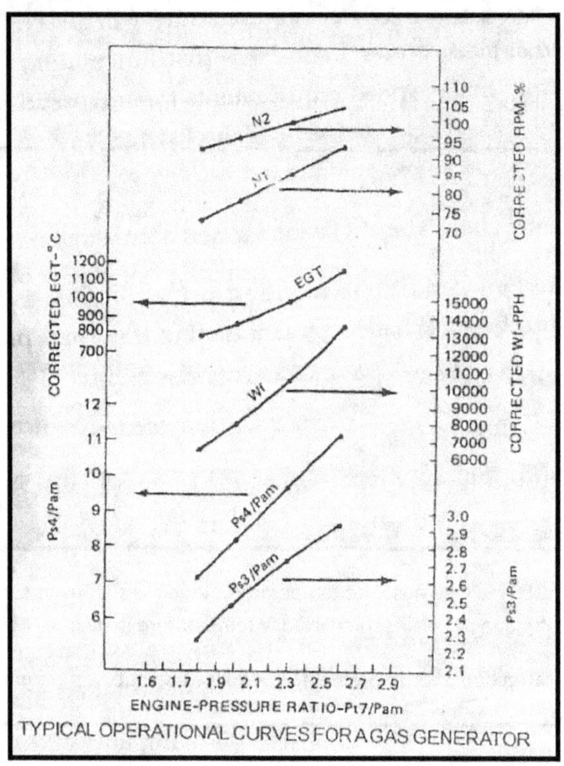

The foregoing discussion is presented as an example of how the performance of an engine can be determined. In actual practice, other tests may be performed and other parameters measured. For any particular type or model of engine, specific instructions are made available by the manufacturer for the testing of the engine in a test cell or on the aircraft.

Operational Checks

To ensure that a gas-turbine engine is in satisfactory operating condition, engine and aircraft manufacturers specify certain operational checks to be routinely performed by maintenance personnel. The particular types of checks and the procedures to be vary, depending on the type of engine and aircraft involved. In this section, to provide an example of typical checks, the checks recommended for the General Electric CF6-50 engine are described briefly.

Dry Motoring Check

The dry motoring check may be required during or after inspection or maintenance to ensure that the engine rotates freely, that instrumentation functions properly, and that starter operation meets speed requirements for successful starts. This check is also used to prime and leak-check the 'lubrication system when maintenance has required replacement of system components.

A dry motoring check should be performed according to the following procedure:

1. Ascertain that all conditions required prior to a nor-mal start are met. These conditions can be established by conducting a normal prestart inspection.

2. Position engine controls and switches as follows:

    a. Ignition, OFF.

    b. Fuel shutoff lever, OFF.

    c. Throttle, IDLE.

    d. Fuel boost, ON.

3. Energize the starter and motor the engine as long as necessary to check instruments for positive indications of engine rotation and oil pressure.

4. De-energize the starter and make the following checks during coastdown:

    a. Listen for unusual noises. Check for roughness. Normal noise consists of clicking of compressor and turbine blades, and gear noise.

    b. Inspect the lubricating system lines, fittings, and accessories for leakage.

    c. Check the oil level in the oil tank.

Wet Motoring Check

When it is necessary to check the operation of fuel-system components after removal and replacement or to perform a depreservation of the fuel system, the wet motoring check is employed. This is accomplished as follows:

1. Position the engine controls and switches as for a dry motoring check.

2. Energize the starter.

3. When core engine speed (N_2) reaches 10 percent, move the fuel shutoff lever to ON and check for oil pressure indication.

4. Continue motoring the engine until the fuel flow is 500 to 600 lb/h or for a maximum of 60 s. Observe the starter operating limits.

5. Move the fuel shutoff lever to OFF and continue motoring the engine for at least 30 s to clear the fuel from the combustion chamber check to see that fuel flow drops to zero.

6. De-energize the starter and, during coastdown, check for unusual noises as mentioned for a dry motoring check.

7. Inspect the fuel system lines, fittings, and accessories for leakage.

8. Check the concentric fuel shroud for leakage. No leakage is permitted.

9. Inspect the lubrication system for leakage.

10. Check the oil level in the oil tank.

Idle Check

The idle check consists of checking for proper engine operation as evidenced by leak-free connections, normal operating noise, and correct indications on engine-related instruments. Engine drain lines must be disconnected from drain cans to check for leakage.

After the engine is started according to approved procedure, the following steps should be taken:

1. Stabilize engine at ground idle.

2. Check fan speed (N_1) core engine speed (N_2), oil pressure, and exhaust-gas temperature (EGT) to see that they are within the proper ranges according to the ground idle speed chart and engine specifications. Engine speeds will vary according to compressor inlet temperature (T_{t2})

3. Visually inspect fuel, lubrication, and pneumatic lines, fittings, and accessories for leakage

4. De-energize flight-idle solenoid. During operations above ground idle, do not exceed the open-cowling limitations imposed by the airframe manufacturer.

5. Stabilize at flight idle and check the same parameters checked for ground idle. See that they are within the limitations set forth on the flight idle speed charts.

Power Assurance Check

The power assurance check is performed to make sure that the engine will achieve takeoff power on a hot day without exceeding rpm and temperature limitations. During the tests, the engine being tested is not used to supply power for any aircraft, system electric, hydraulic, or other. The engine is tested at 50 percent, 75 percent, and maximum power.

During engine operation for the power assurance check, EGT must be observed constantly to avoid the possibility of over temperature. Should the temperature approach maximum allowable, the throttle must be retarded sufficiently to hold the EGT within limits. In the operation of the engine, the throttle should always be moved slowly.

To perform the power assurance test, follow these steps:

1. Set the engine power at nominal N^2 speed as indicated on the appropriate chart for the total air temperature (TAT). For example, the nominal N^2 curve on the chart may coincide with the 91.8 percent line at 10°C for the 50 percent power setting. The throttle will therefore be adjusted to produce 91.8 percent rpm when the TAT is 10°C for a 50 percent power setting.

2. Four minutes after the throttle lever is set, record the average readings of TAT, N_1 speed, N_2 speed, EGT, EPR (engine pressure ratio), and fuel flow (W_f). Correct W_f for local barometric pressure in accordance with instructions.

3. Using N_1 (where N_1 = target N_1 - observed N_1) as a correction factor, adjust readings according to the parameter adjustments set forth in the operations manual.

In the operation of gas-turbine engines of any type, it must be emphasized that temperatures and rpm for both N_1 and N_2 must be watched carefully. If it is expected that a beyond-limits condition is developing, the operator should take immediate action by retarding the throttle or shutting the engine down.

Before a hot engine is shut down, it should be operated at ground idle speed for about 3 min to permit temperature reduction and stabilization. As soon as the engine is shut down, the EGT gage should be observed to see that EGT starts to decrease. If EGT does' not decrease, an internal fire is indicated, and the engine should be dry-motored at once to blowout the fire.

During coastdown after the engine is shut down, a technician should listen for unusual noises in the engine such as scraping, grinding, bumping, and squealing.

Fuel Control Adjustments

Fuel control maintenance on the flight line is usually limited to removal and replacement of the fuel control unit and resetting of the control adjustments. The adjustments include specific gravity, idle RPM and maximum power on hydro-mechanical or hydro-pneumatic controls. On engines with full schedule electronic engine controls, or with FADECs, few if any adjustments are necessary because the system adjusts and compensates on its own.

    a.   **Specific Gravity Adjustment**

When a run-up is to be made for gas turbine engine performance checks, the normal fuel prescribed by the manufacturer should be used because the Btu value, as well as specific gravity, may be different in an alternate fuel and the validity of the check could be compromised.

Otherwise a specific gravity adjustment can be made. It is a means of resetting tension on the Differential Pressure Regulating Valve Spring within the fuel control when an alternate fuel is being used and the engine is undergoing performance testing.

b. Performance Checks (Turbojet and Turbofan Engines)

1. Acceleration Check

In conjunction with the trim procedure, an acceleration check is usually accomplished as an additional test of engine performance. After the trim check is completed, a mark is placed on the cockpit power lever quadrant at the takeoff position. The power lever is then advanced from idle position to takeoff position, and the time is measured against a published tolerance. The tune, even for a large gas turbine engine, is quite low, in the range of five to ten seconds.

2. Throttle Spring-back and Cushion Checks

Another important part of the trim procedure is to check for power lever spring-back and cushion. The operator, before and after the trim run, moves the power lever full forward and releases it. The distance the lever springs back is measured against prescribed tolerances. On an airliner for instance, this distance might measure one-quarter inch or more. When spring-back is correct, the fuel control will reach its internal stops before the cockpit quadrant reaches its forward stop. If out of limits, an adjustment of the aircraft control system which attaches to the fuel control is required by the technician.

Cushion is the distance in inches from the power lever at take-off setting to full power lever travel.

The cushion check insures that the pilot will not only be able to obtain takeoff power but also additional power lever travel in case of emergencies.

c. Part Power Trim

In order to save wear on the engine and also to save fuel, both engine pressure ratio rated engines and speed rated engines are generally trimmed at less than takeoff power. This procedure involves closing all service bleeds to avoid compression loss and placing a physical obstruction in the path of the fuel control lever linkage called a part power trim stop. The power lever is advanced to hit the stop during trimming and the trimming adjustment is made at this position with the engine operating at approximately the "maximum continuous" thrust rating. To stabilize internal fuel control linkages, trim adjustments are made in the increase direction. After trimming, the stop is removed in order to perform a takeoff power check.

d. Two Types of Trim Procedures-Engine Pressure Ratio Trim and Speed Trim

Thrust producing turbine engines utilize either the engine pressure ratio trim or the fan speed trim procedure. If the engine is configured with an engine pressure ratio system, the pilot will use a cockpit engine pressure ratio gauge to set engine power. In terms of trimming, the engine is referred to as an engine pressure ratio rated engine. If the engine does not have an engine pressure ratio system, it is trimmed in accordance with fan speed and the pilot in this case uses a tachometer indicator to set engine power. In terms of trimming, this engine is referred to as a fan speed rated engine.

Power checks for turboprop and turboshaft engines are discussed later in this Chapter.

1.   Engine Pressure Ratio (EPR)

The aircraft utilizes an engine pressure ratio gauge for thrust indication. This gauge conveniently reads the ratio of turbine discharge pressure divided by compressor inlet pressure and automatically corrects for changing ambient conditions.

EXAMPLE: If turbine discharge pressure is measured to be 58.2 inches of mercury when the engine is correctly trimmed and engine inlet pressure reads 30.0 inches of mercury, the engine pressure ratio is calculated as follows:

within the accepted tolerance of 1.94, both the engine pressure ratio indicating system and the engine are performing satisfactorily.

2.   Trimming Procedure for Engine Pressure Ratio Rated Engines

On many axial flow engines, there exists a better relationship between internal engine pressures and thrust than engine RPM and thrust. In fact, on some dual-spool engines the last ten percent of the RPM range can increase thrust by as much as 30 percent. On these engines, engine pressure ratio is a quicker measure of thrust than by use of RPM/ thrust charts.

By turning the maximum trim adjustment the fuel flow and thrust will be increased or decreased. This in turn will affect the relationship between compressor inlet pressure and turbine discharge pressure. That is, as the latter goes up with fuel flow increase, engine pressure ratio will also go up for a given compressor inlet condition.

The trim procedure is as follows:

With finely calibrated gauges, maintenance personnel measure inlet conditions f barometric pressure and ambient temperature before running the engine. They will insert the part power trim stop, and install the necessary fittings to measure turbine discharge pressure. Finally, with the engine running on the part power stop, they will trim the engine to a value prescribed by a manufacturer's graph.

Cautions to be observed at this time are: Do not use the temperature given at the control tower; there could be a measurable difference between that reading and the on-site reading. Also, do not use the aircraft Outside Air Temperature (OAT) gauge; it could be heat soaked from the sun and again be significantly different from a thermometer reading taken at the engine. An industry wide procedure is to hang the thermometer in the shade of the nose wheel-well until the temperature stabilizes.

Manufacturers also commonly recommend that two or more accelerations be made to a high power setting to properly pre-load the internal mechanisms within the fuel control and ensure a higher degree of repeatability before the actual acceleration to trim power setting is made.

After calculating the necessary turbine discharge pressure for an outside air temperature of 75°F and inlet pressure of 30 inches of mercury. If the engine is not producing 58.20 inches of mercury turbine discharge pressure, maintenance personnel will up-trim the engine to this value.

At that time, maintenance personnel will mathematically compute 58.2 inches of mercury turbine discharge pressure divided by 30.0 inches of mercury compressor inlet pressure, to be an engine pressure ratio of 1.94. The cockpit gauge must at that time read 1.94, plus or minus a typical tolerance value. If it does not, the engine pressure ratio indicating system must be checked.

a.   **Full Power Check after Trimming**

After the part power trim is completed, the part power stop is removed and the power lever is advanced toward the full power Turbine Discharge Pressure value of 67.0 inches of mercury. If this value can be achieved without exceeding the limits of engine condition instruments, such as engine speed exhaust gas temperature, and fuel-flow, the engine full lower check is satisfactory. This check is further discussed in the following paragraphs on trimming of speed rated engines.

3.   **Trimming Procedure for Speed-rated Engine**

A fan speed trim is, generally performed on turbofans of the dual-spool configuration where there exists a direct relationship between fan speed (N1) and thrust. Counterclockwise rotation of the maximum adjustment will increase fuel flow to the engine, increase fan speed, and, consequently, engine thrust output. Adjusting the idle set screw will reset idle speed to the manufacturer's recommended percent RPM value. The amount of adjustment will depend on the parameters of an engine performance curve.

The fan speed trim determination in this case is dependent on ambient temperatures. The example engine is speed rated. That is, its thrust is determined by a comparison to fan speed. High pressure compressor speed, turbine temperature, and other engine operational parameters must fall within an allowable range when rated fan speed is obtained. If they do not, one of two conditions probably exists, a turbo-machinery malfunction or a fuel scheduling malfunction as follows:

a)   Dirty compressors and damaged compressors can cause aerodynamic problems. Damaged hot sections can cause thermodynamic problems.

b)   Poor fuel scheduling can cause compressor stalls, over temperatures, and flameout.

In order to obtain required thrust, some part of the engine is being over-taxed, and this shows up on one of the condition monitoring instruments in the cockpit, i.e., RPM, exhaust gas temperature, fuel flow, etc. The cause of the high indicator reading must then be determined by troubleshooting procedures and corrected before the engine is released for service.

To ensure accuracy of readings, many operators utilize a portable precision tachometer slaved into the aircraft percent rpm indicating system. This gives the trim crew personnel an immediate indication of condition of the cockpit instrument and more information with which to completely assess the condition of the engine.

The trim procedure is as follows :

1.   With a precision thermometer reading taken at the aircraft, plot the fan (N1) speed for the ambient temperature.

2.   Deploy trim part power stop;

3.   Run aircraft, advancing power lever until linkage hits trim stop;

4.   Adjust maximum setting and record gauge readings check against manufacturer's limits;

5.   Retract trim stop and advance power lever to N1 value. Check N2 speed and Ts to be within limits.

   c.   Data Plate Speed Check

A performance check that is quite often completed along with the trim check is the data plate speed check. On many gas turbine engines a small metal plate is attached at time of manufacture during final performance testing.

The plate is stamped with the engine speed at which required thrust value at part power trim was obtained. No two engines would necessarily be stamped with the same speed because production tolerances of engine parts vary the speed to thrust relationship on nearly every engine in some measurable way. However, all engines produced will probably be within 2 to 3 percent RPM of each other.

The purpose of this check is to compare future engine performance against the "as new" performance data. For example, consider a hypothetical data plate is stamped (87.25 percent N2 Speed at 1.61 engine pressure ratio, 59°F). As a ground performance check on engine condition when the engine accumulates service time (cycles and hours), the data plate speed check can be accomplished. The procedure in this instance would be to operate the engine at 1.61 engine pressure ratio and observe the N2 tachometer indication, then, compare that indication to the "as new" speed on the data plate. If the tolerance is (+2.0 percent) and the engine being tested goes to 89.5 percent N2 speed, the engine is out of limits for the data plate speed check because the limit is 89.25 percent.

This check is an assessment of engine performance and not necessarily a pass or fail type check. If the engine is otherwise performing satisfactorily, the operator might decide to keep the engine in service, even though it is no doubt consuming more fuel to give the required 1.61 engine pressure ratio than when new.

It might also tell the operator that the compressor is contaminated and requires field cleaning or that the hot section is deteriorating and should be scheduled for repair at the next inspection interval. The test run will seldom, if ever, be accomplished on a Standard Day 59°F, so the observed speed on the tachometer indicator will have to be corrected as per manufacturers charts before comparison to the data plate speed.

The discussion above described an engine pressure ratio rated engine. But the data plate speed check is equally applicable to the turbofan speed rated engine, or the turboprop engine, which is rated in torque units or horsepower units.

f.   Trim Restrictions

There are always ambient condition restrictions on trimming which must be closely followed. Excessive wind in the direction of the tailpipe will cause a false high turbine discharge pressure and subsequent low trim when the wind is later within limits.

Excessive wind in the inlet will cause a false high compression and turbine discharge pressure and a subsequent low trim when wind conditions drop within the limits.

These situations occur as follows:

1.   Wind up the tailpipe causes a back pressure and a false high turbine discharge pressure for which the technician would compensate by down-trimming. Then, later, in calm air, turbine discharge pressure would be low and, consequently, engine pressure ratio would be low;

2.   Wind into the inlet causes a false high compressor inlet pressure which generally affects compression to a greater degree than it affects the cockpit engine pressure ratio reading.

The compressor magnifies the false high inlet pressure by increased compression which affects a rise in turbine discharge pressure. This causes what appears to be a higher engine pressure ratio reading or over-trimmed engine. The trim crew will compensate for this by, down-trimming and then later in calm air the engine will be under-trimmed. Wind is also seldom steady and, when gusting, it causes erratic gauge readings which make trimming calculations difficult to obtain.

Other trim restrictions such as moisture content (rain) and moisture (icing) will also be found in the aircraft operations manual and must be followed or a false trim will result.

The same restrictions apply to the speed rated engines.

Even though the trim instrumentation is not as directly affected, the engine performance will be affected, and an incorrect trim could result.

g.   Trim Danger Zones

An inherent danger to personnel and equipment exists in the area of operating gas turbine engines. All flight line personnel must be constantly alert to keep out of the danger areas when approaching or leaving the proximity of an operating engine. Loose articles of clothing, microphone chords, mechanic wipe rags, tools, etc., must be carefully secured so they will not beingested bby the engine. For maximum safety use an intercom system between the cockpit and the ground crew to minimize or use hand signals (included in Appendix 5 of this book).

h.   Ear Protection from Noise

A large turbojet or turbofan engine operating at full power settings can generate sound levels up to 160 decibels at the aircraft. Smaller engines of all types can generate sound levels up to approximately 130 decibels. This noise intensity is sufficient to cause either temporary or permanent hearing loss if adequate ear protection is not used by ground personnel.

The most effective ear protection device is the muff type, which fits over the entire ear and defends against noise to the ear opening and also to the bone structure behind the ear. Ear plugs provide minimal hearing loss protection and are generally only recommended for use in lower noise areas or for low time exposure to higher noise areas.

i.   Flat Rating

Today, most gas turbine engines are characterized as flat-rated. This refers to the flat shape of the full power curve and the point on the ambient temperature scale at which the power starts to drop below 100 percent. Analysis of the curve will reveal that a fan speed of 96 percent RPM corresponds to 100 percent thrust on this engine, and thus this value can be obtained at any ambient temperature upto 90°F. That is, by moving the power lever more and more forward, the pilot can obtain rated thrust at any temperature up to 90°F. After 90°F, more forward movement of the power lever is' not permitted because it most: likely will result in an engine over temperature.

When ambient temperature exceeds the flat rating of the engine, 100 percent thrust can no longer be obtained. This being the case, the aircraft's gross weight might need to be adjusted, or at the very least, runway takeoff roll will increase and the flight crew will need to account for this in their take-off procedures.

Some engines are flat rated to only 59°F, others over 100°F. This consideration depends largely on the needs of the aircraft manufacturer. Generally, fla' rating is believed to enable the engine to produce a constant rated thrust over a wide range of ambient temperatures without working the engine harder than necessary, in the interest of prolonging engine service life.

For example, an engine rated at 3,500 pounds thrust at 59°F might be re-rated to 3,350 pounds thrust at 90°F. The aircraft user might riot need to utilize 3,500 pounds thrust, nor the maximum gross weight of the aircraft, and he would like to benefit from increased engine service life and lower fuel consumption by operating at 3,350 pounds thrust maximum. Flat rating is an engine manufacturer's way of re-rating an engine to a lower rated thrust than it would have 'at Standard Day temperature. The engine will be able to use that lower rated thrust over a wide ambient temperature range. Flat-rating is equally applicable to all types of gas turbine engines, both thrust producing engines and torque producing engines. The aircraft manufacturer will probably use the following process, or one very similar, when selecting the flat-rating that best suits his needs.

1.   The user decides the take-off power needed for his aircraft configuration, route requirements, runway lengths, runway altitudes, etc.

2.   The user calculates the highest ambient temperature at which required takeoff power can be obtained.

3.   The engine and aircraft manufacturer print all of the flight manuals, operational instructions, etc., to reflect the selected takeoff power as the maximum usable for normal operation.

From this example, it can be seen that if an engine is re-rated to a lower power level than it is capable of producing, the engine still retains its full capability of power as a reserve for emergencies. It can also be seen that no mechanical changes are needed to the engine or fuel scheduling system, merely changes in the printed operational data.

i.   Turboprop and turboshaft engine performance checks

Turboprop and turboshaft engines are torque producers rather than thrust producers. Performance calculations of torque producing engines are very different from that of turbofan and turbojet engines which use engine pressure ratio or fan speed as power indicators.

The turboprop and turboshaft aircraft cockpit contains a torque indicator that reads out in either torque (psi) oil pressure, torque (ft/lbs.), or torque (%). The gauge reading is produced most often by either a "phased shift" electronic torquemeter system or a "balanced piston" torquemeter oil system.

To maintain the installed engine on aircraft and engine in storage in serviceable condition, the periodical inspection is essential.

The following schedule inspection to be complied on engine.

1.	Daily inspection	2.	walk around inspection
3.	lay over inspection	4.	check A,B,C and D
5.	hot section inspection	6.	heavy maintenance inspection
7.	overhaul of an engine		

Engine in Storage

Carry out inhibition for short term or long term storage

Preparing Engine for Storage and Transportation

The preparation of the engine for storage and transportation is of major importance, because storage and transportation call for special treatment to preserve the engine. So that the fuel system will resist corrosion during storage, it is filled with a special oil and all openings are sealed off. The external and internal surfaces of the engine are also protected, either by special inhibiting powders or by paper impregnated with inhibiting powder. The engine is enclosed in a reusable bag or plastic sheeting into which a specific amount of desiccant is inserted. If the engine is to be transported, it is often packed in a crate or metal case.

Engine Trimming and Adjustment

Trimming a gas-t-turbine engine is the process of adjusting the fuel control unit so that the engine will produce its rated thrust at the designated rpm. The thrust is determined by measuring the engine pressure ratio (EPR), which is the ratio of turbine discharge pressure to engine inlet pressure (Pt6/Pt2). On engines equipped with variable compressor vanes, it is necessary to check the vane angles and the operation of the engine vane control (EVC) during the trimming process. The trimming of a gas-turbine engine may be compared with the tuning of a piston engine for optimum performance Gas-turbine engines with computer-controlled fuel systems do not require trimming because trimming adjustments are made automatically by the fuel control computer.

Gas-turbine engines manufactured by Pratt & Whitney are tested at the factory and adjusted to produce rated thrust. The engine speed (N_2) which is required for the engine to deliver, rated thrust is stamped on the engine data plate or recorded on the engine data sheet of the engine log book. This information is supplied in both rpm and percent of maximum rpm. Because of manufacturing tolerances and slight variations which occur during the manufacture of engines, no two engines are exactly alike, and very rarely will two engines of the same model produce rated thrust at exactly the same rpm. The rpm for rated thrust stamped on the data plate will therefore vary from engine to engine.

Engine trimming is required from time to time because of changes that take place during the life of the engine. Dust and other particulate matter will adhere to the surfaces of the compressor rotor blades and stator vanes and lead to a slight resistance to airflow. Erosion of the leading edges of blades and vanes. caused by dust, sand, and other material changes the characteristics and performance of the compressor. The turbine blades and vanes, which are exposed to very high temperatures, are subject to corrosion, erosion, and distortion. All the foregoing factors tend to cause the engine thrust to decrease over a period of time; therefore, trimming is necessary to restore the rated performance of the engine. Generally speaking, when an engine indicates high exhaust-gas temperature (EGT) for a particular EPR, it means that the engine is out of trim.

The following general principles for trimming are for information only:

1.   Head the airplane as nearly as possible into the wind. Wind velocity should not be more than 20 mph for best results. See that the area around the aircraft is clean and free from items which could enter the engine or cause other problems during the engine run.

2.	Install the calibrated instruments required for trimming. One of these instruments is a pressure gage for reading turbine discharge pressure (P_0) or EPR. Another important instrument is the calibrated tachometer, which is used to read N_2 rpm.

3.	Install a part-throttle stop or fuel control trim stop as specified in the trim instructions.

4.	Record ambient temperature and barometric pressure. These values are necessary to correct performance readings to standard sea-level conditions. The pressure and temperature information is used to determine the desired turbine discharge pressure or EPR by means of the trim curve published for the engine.

5.	Start the engine and operate it at idle speed for the time specified to ensure that all engine parameters have stabilized. Operate the engine at trim speed as established by the trim stop on the fuel control for about 5 min to stabilize all conditions. The overboard air-bleed valves should be fully closed and all accessory air bleed must be turned off.

6.	Observe and record the P_o or EPR to determine how much trimming (if any) is required. If trimming is required, adjust the fuel control unit to give the desired Po or EPR. When this is attained, record the engine rpm, the EGT, and the fuel flow.

7.	The observed rpm is corrected for speed bias by means of a temperature-vs.-rpm curve to provide a new engine trim speed in percent corrected to standard conditions.

Note that these procedures will vary considerably, depending on the type and model of engine being trimmed. The purpose of trimming for all engines, however, remains the same: to provide optimum engine performance without exceeding the limits of rpm and temperature established for the engine. As explained previously, engines equipped with computer-controlled fuel controls do not require periodic trimming because the adjustments are made automatically by the computer.

As mentioned previously, a variety of detectors, sensors, instruments, and systems have been developed to detect incipient and existing faults and to provide information regarding the adjustment and calibration of operating units. Some systems are installed aboard the aircraft so that fault indications can be obtained during flight, and others are designed as ground support equipment. One ground support system is the Jetcal Analyzer/Trimmer manufactured by Howell Instruments, Incorporated, Fort Worth, Texas.

Description

The Jetcal analyzer consists of an instrument case and an accessory case mounted on wheels for mobility. By means of the case handle, the unit can be easily moved about.

Housed in the instrument case is a probe controller assembly and a portable trimmer assembly. The trimmer assembly may be removed from the instrument case to be used as a separate instrument in the aircraft cockpit during trimming procedures. When the trimmer is used separately from the analyzer, the interconnect cable is disconnected and the trimmer is powered by the analyzer power cable and a power-cable adapter. Cables, cable adapters, and heater probes are stored in the accessory case.

The Jetcal analyzer is used (1) to check continuity of the EGT system, insulation resistance of the EGT system, resistance of thermocouple harnesses, accuracy of engine thermocouples, engine thermocouple temperature spread during engine trim, accuracy of the EPR system, accuracy of aircraft rpm systems that use standard tachometer generators, and calibration of the aircraft EGT indicator; (2) to bench-test individual thermocouples; (3) to test overheat detection systems; (4) to monitor EGT and rpm during engine trim and pressure and EPR during engine trim; and (5) to correct temperature and rpm readings to standard-day conditions. The standard-day corrections generated by the analyzer make it unnecessary to convert the engine parameters to standard-day conditions mathematically or by means of charts.

The instrument and control panels for the Jetcal Analyzer Trimmer are shown is figure. Note that the temperature, EPR, and rpm information is presented in digital displays.

Except for engine-trimming operations, the Jetcal analyzer can perform its functions without the necessity of running the engine. This, of course, saves both time and fuel in the routine maintenance and troubleshooting of the engine systems.

## Chapter - 10

FIRE PROTECTION

FIRE EXTINGUISHING SYSTEMS

Before understand how fires are extinguished, everybody should know what makes a fire burn. A fire is simply a chemical reaction that occurs when oxygen combines with a fuel to produce heat and, in most cases, light. Three elements must be present for a fire to occur, including:

1. A combustible fuel

2. A supply of oxygen

3. Heat.

If you remove anyone of these three elements, combustion cannot be sustained.

The easiest elements for a fire extinguishing agent to remove are the oxygen supply and the heat required to start combustion. Therefore, fire extinguishing systems are designed to dilute oxygen levels to a point that does not support combustion or to reduce the temperature below the ignition point.

Classification of Fires

All fires are classified by the National Fire. Protection Association (NFPA) according to the type of combustible fuel involved. For example, a **Class A fire** is one in which solid combustible materials such as wood, paper, or cloth burn. An aircraft cabin fire is a good example of a Class A fire.

Class B fires, on the other hand, involve combustible liquids such as gasoline, oil, turbine fuel, hydraulic fluid, and many of the solvents used in aviation maintenance. Class B fires are the most common type of fire encountered in an engine nacelle.

Class C fires are those which involve energized electrical equipment. Special care must be exercised when trying to extinguish a Class C fire because of the dangers presented by both 'the electricity and the fire itself.

Class D fires involve a burning metal, such as magnesium, and burn extremely hot. Because Class 0 fires burn so hot, the use of water or other liquids on Class 0 fires causes the fire to burn more violently or explode.

Engine Fire Zones

The powerplant area is divided into fire zones based on the volume and smoothness of the airflow through the engine compartment. These classifications allow manufacturers to match the type of detection and extinguishing system to the fire conditions. Do not confuse these classifications with the NFPA fire classifications discussed earlier.

Class A fire zones have large quantities of air flowing past regular arrangements of similarly shaped obstructions. The power section of a reciprocating engine where the air flows over the cylinders is an example of a Class A fire zone.

Class B fire zones have large quantities of air flowing past aerodynamically clean obstructions. Heat exchanger ducts and exhaust manifold shrouds constitute Class B fire zones. Additional Class B fire zones include cowlings or tight enclosures that are smooth, free of pockets, and adequately drained so leaking flammables cannot puddle. Turbine engine surfaces sometimes fall within this class if the engine's surfaces are aerodynamically clean and all airframe structural formers are covered by a fire-proof liner to produce a smooth enclosure.

Class C fire zones have relatively small quantities of air flowing through them. The compartment behind the firewall is considered to be a Class C fire zone.

Class D fire zones are areas that have little or no air-flow. Wheel wells and the inside of a wing structure are typical Class D fire zones.

Class X fires zones have large volumes of air flowing through them at an irregular rate. Because of the sporadic airflow, Class X fire zones are the most difficult to protect from fire. In fact, the amount of extinguishing agent required to adequately protect a Class X fire zone is normally twice that required for other zones. Class X fire zones are common in engine nacelles.

Troubleshooting

Intermittent alarms or false alarms are probably the most common problems associated with a fire detection system. Most intermittent alarms are caused by an intermittent Short circuit in the detector system wiring. Electrical shorts are often caused by a loose wire that occasionally touches a nearby terminal, a frayed wire brushing against a structure, or a sensing element that has rubbed against a structural member long enough to wear through the insulation. Intermittent faults can often be located by applying power to the system and moving wires to recreate the short.

False alarms can typically be located by disconnecting the engine sensing loop from the aircraft wiring. If the false alarm continues, a short must exist between the loop connections and the control unit. However, if the false alarm ceases when the engine sensing loop is disconnected, the fault is in the disconnected sensing loop. The loop should be examined to verify that no portion of the sensing element is touching the hot engine. If there is no contact, the shorted section can be located by isolating and disconnecting elements consecutively around the entire loop. Kinks and sharp bends in the sensing element can cause an internal wire to short intermittently to the outer tubing. The fault can be located by checking the sensing element with a megohm meter, or megger, while tapping the element in the suspected area to produce the short.

Moisture in the detection system seldom causes a false fire alarm. However, if moisture does cause an alarm, the warning will persist until the contamination is removed or boils away and the resistance of the loop returns to its normal value.

Another problem you could encounter is the failure to obtain an alarm signal when the test switch is actuated. If this occurs, the problem could be caused by a defective test switch or control unit, the lack of electrical power, an inoperative indicator light, or an opening in the sensing element or connecting wiring. Kidde and Fenwal continuous-loop detectors will not test if a sensing element is shorted or broken; however, they will pre-Ade a fire warning if a real fire exists. When the test switch fails to provide an alarm, the continuity of a two-wire sensing loop can be determined by opening the loop and measuring the resistance of each wire. In a single-wire continuous-loop system, the center conductor should be grounded.

Introduction

All gas turbine engines and their associated installations systems incorporate features that reduce the possibility of an engine fire. It is essential, however, that if a failure takes place and results in a fire, there is provision for the immediate detection and rapid extinction of the fire, and for the prevention of it spreading. The detection and the extinguishing systems must add as little weight to the installation as possible.

Supersonic aircraft present special problems of fire detection and extinguishing duo to the increased risk of spontaneous combustion occurring as the power plant temperature rises.

Power plants, used only for lift operations (Part 21) , are essentially lightweight installations and fire protection system is must take account of this. Because the engine is operated for very short periods, a fire detection system only is provided and the intensity of any fire is minimized by having no power plant ventilation. The fire will therefore be extinguished when the fuel is shut off and rotation of the engine ceases.

PREVENTION OF IGNITION

The following features ensure that the prevention of ignition is achieved as far as possible, and in most instances a dual failure is necessary before a fire can occur.

All the potential sources of inflammable fluids are, isolated from the 'hot end' of the engine. External-fuel and oil system components and their associated pipes are usually located around the compressor, casings, in a 'cool' zone, and separated by a fireproof, bulkhead from the combustion, turbine and jet pipe area, or 'hot' zone. The zones may be ventilated, as described in Part 13, to prevent the accumulation of inflammable vapours.

All pipes that carry fuel, oil, or hydraulic fluid, are made fire resistant, and all electrical components and connections are made flame-proof. Sparking caused by discharges of static electricity is prevented by bonding all aircraft and engine components; this gives electrical continuity between all the components and makes them incapable of igniting an inflammable vapour.

The power plant cowlings are provided with an adequate drainage system to remove inflammable fluids from the nacelle, bay, or pod, and, all seal leakages from components are drained -overboard.

Spontaneous combustion can be prevented on aircraft flying at high Mach numbers by ducting boundary layer bleed air around the engine. However, if ignition should occur, this high velocity air stream may have to be shut off, otherwise it would increase the flame intensity and reduce the effectiveness of the extinguishing system by rapid dispersement of the extinguishant.

FIRE DETECTION

The rapid detection of a fire is essential so that the necessary extinguish drill can be effected before the fire becomes too large. It. is also extremely important that a fire detection system will not give a false fire warning due to short circuiting, which can result from vibration, chafing or the ingress of moisture. Two separate fire detecting circuits, installed side by side within the fire warning system, may be used to obviate a false fire warning signal.

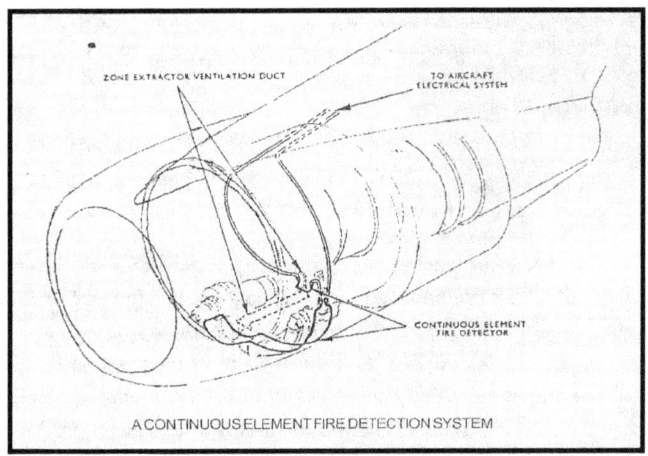

A CONTINUOUS ELEMENT FIRE DETECTION SYSTEM

A detection system may consist of a number of strategically located detector units, or be of the continuous element sensing type that can be shaped and attached to preformed support tubes. The sensing element can be routed across outlet orifices, such as a zone extractor ventilation duct, to give early detection of fire. The presence of a fire is signaled by a change in the electrical impedance or the output voltage of the detector circuit, according to the type of detector, thermistor or thermocouple. The change in temperature creates the signal which, through an amplifier, operates the warning indicator. Fire indication is given by a light or bell, and the warning is cancelled when the detector cools after the fire has been extinguished.

At aircraft speeds above Mach 3, the considerably higher temperature levels may be such as to render the thermistor or thermocouple fire detection system unsatisfactory. Thermal detectors that sense either a temperature rise, or a rate of temperature rise, may therefore prove most suitable.

Alternatives to the above types are surveillance detectors that respond to light radiation from a fire. These may be made so sensitive that they respond only to the ultra-violet rays emitted from a kerosene fire.

FIRE CONTAINMENT

An engine fire must be contained within the power plant and not be allowed to spread to other parts of the aircraft. The cowlings that surround the engine are usually made of light alloys, which would be unable to contain a fire when the aircraft is static. During flight, however, the airflow around the cowlings provides sufficient cooling to render them fireproof. Fireproof bulkheads and any cowlings that are not affected by a cooling airflow, and sections of cowlings around certain outlets that may act as 'flame holders' are manufactured from steel or titanium,

FIRE EXTINGUISHING

Before a fire extinguishing system is operated, the engine must be stopped to reduce the discharge, of inflammable fluids and air into the fire area. Any valves, such as the low pressure fuel cock, that control the flow of an inflammable fluid must be situated outside the 'hot' zone to prevent fire damage rendering them inoperative.

After a fire has been extinguished, no attempt must be made to start the engine again as this would probably re-establish the fluid leak and the ignition source that were the original causes of the fire. Furthermore, the extinguishing system may be exhausted.

The extinguishant that is used for engine fires is usually methyl-bromide or one of the Freon compounds. Pressurized containers are provided for the extinguishant and these are located outside the fire risk zone. When the, relevant electrical circuit is manually operated, the extinguishant is discharged from the containers through a series of perforated spray pipes or nozzles into the fire. The discharge must be sufficient to give a predetermined concentration of extinguishant for a period that may vary between 0.5 seconds and 2 seconds. System generally enables two separate discharges to be made.

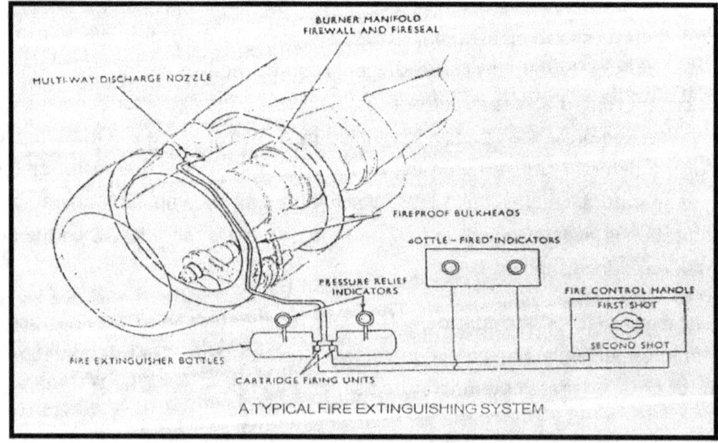

ENGINE OVERHEAT DETECTION

Overheating or the "hot end" of the engine does not constitute a serious fire risk. Detection or an overheat condition, however, is essential to enable the pilot to stop the engine before mechanical or material damage results.

A warning system of a similar type to the fire detection system, or thermocouples suitably positioned in the cooling airflow, may be used to detect excessive temperatures. Thermal switches positioned in the engine overboard air vents, such as the cooling' air outlets, may also be included to give an additional warning.

## Chapter -11

PROPELLER INSPECTION, MAINTENANCE AND INSTALLATION

As an aviation maintenance engineer, it will be required to inspect and perform routine maintenance on various types of propellers. Therefore, in addition to being familiar with propeller maintenance procedures, everybody must have a basic knowledge of maintenance regulations that relate to propellers. It is important to note that as a certified mechanic with a powerplant rating, you are authorized to perform only minor repairs and minor alternations to propellers.

Maintenance Regulations

FAR Part 43 spells out, Maintenance, Preventative Maintenance, Rebuilding, and Alteration, defines the different classes of maintenance for propeller systems. Appendix D lists the minimum requirements for 100-hour and annual inspections of propellers and their controls. During an annual or 100-hour inspection, propeller assemblies must be checked for cracks, nicks, and oil leakage. In addition, bolts must be inspected for proper torque and appropriate safe tying. When equipped with anti-icing devices, those devices must be inspected for improper operations and obvious defects. Also, it is required to inspect propeller control mechanisms for proper operation, secure of mounting, and restricted travel.

Appendix A of FAR 43 lists propeller major alterations and repairs which must be performed by the manufacturer or a certified repair station. Propeller major alterations include changes in blade design, hub design, and governor or control design. Also included are installations of a propeller governors, feathering systems, de-icing systems, and part not approved for the propeller. On the other hand, propeller major repairs include items such as retipping, replacement of fabric covering, and inlay work on wood propellers. Also included in fixed-pitch wood propeller major repairs are replacement of outer laminations and repair of elongated bolt holes in the hub. Any repairs to, or straightening of, steel propeller blades is considered to be a major repair. In addition, repairs to, or machining of, steel hubs is considered to be a major repair. Major repairs to aluminium propellers include shortening or straightening of blades and repairs to deep dents, cuts, scars, and nicks.

Other major repairs listed in Appendix A of Part 43 included the repair on replacement of internal blade elements such as internal de-icer heating elements. Also listed are controllable pitch propeller overhauls and repair of propeller governors. Given the number of items that are considered to be either major repairs or alterations to propellers, it may be difficult to differentiate whether a specific repair or alteration is minor or major. Therefore, if you are in doubt about the status of a contemplated alteration or repair, you should contact your local Flight Standards Office of the FAA.

Inspection and Maintenance

Specific inspection items and minor maintenance tasks for which you are responsible depend on the type of propeller and its accessories. The following provides generic information on typical inspection and maintenance procedures. However, the information provided here is only general in nature and, therefore, you should always consult the appropriate aircraft or propeller maintenance manuals and service bulletins for specific instructions and service limits.

Wood Propellers

A fixed-pitch wood propeller is simple in concept and operation; however, the fine details of its construction require close visual inspection. For example, both annual and 100-hour inspection require to check for cracks, nicks, and properly torqued or safetied bolts. While required visual inspections are mostly conducted with the propeller mounted on the engine, there may be occasions when removal is necessary. For example, if excessive vibration exists, you may have to remove the propeller to check for elongated mounting bolt holes or proper balance.

To facilitate an inspection, a propeller should be cleaned. Wood propellers may be cleaned with warm water and a mild soap, using brushes or a cloth. If the aircraft operates near salt water, the propeller should be flushed with fresh water often. If a visual inspection after cleaning reveals defects which must be further examined or repaired, propeller removal may be necessary. Removal of a wood propeller is usually a simple matter of removing a spinner, safety devices such as cotter pins or wire, and the mounting bolts. In all instances, follow the recommended removal procedures outlined in the maintenance instructions for the aircraft and engine.

Common defects found in wood propellers include separation of the laminations and dents or bruises on the surface. Other possible damage includes cracks or scars across the blade back or face, broken sections, warping, and worn or oversize center bore and bolt holes. If a dent, bruise, or scar is found on a. blade surface, inspect the damage with a magnifying glass while flexing the blade to help expose any cracks.

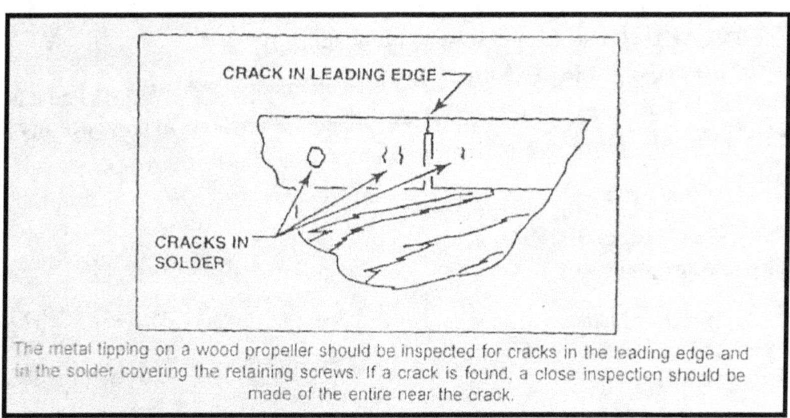

The metal tipping on a wood propeller should be inspected for cracks in the leading edge and in the solder covering the retaining screws. If a crack is found, a close inspection should be made of the entire near the crack.

When inspecting metal tipping, look for looseness or slipping, loose screws or rivets, and cracks in the solder joints. If a crack is found in a solder joint near the blade tip, it may be an indication of wood deterioration. Therefore, the area near the crack should be inspected closely while flexing the blade tip. If no defects are found, the joint may be resoldered; however, the blade tip should be inspected at closer intervals for a recurrence of cracking.

When inspecting the wooden blades installed on a ground-adjustable propeller, special attention should be given to the metal sleeve and shank area. The presence of cracks in these areas may indicate broken or loose lag screws.

Repairs

Typically, small, cracks parallel to the grain or small cuts on a wood propeller may be repaired by working resin glue into the crack. Once the glue is dry, the area is sanded with fine sandpaper and refinished with an approved varnish or other coating. Repairable tip fabric defects include cracks, bubbles, paint chipping, and wrinkles that appear when the tip is twisted or flexed. If the tip fabric has surface defects of 3/ 4 inch or less, and a breakdown in the wood structure is not suspected, the defect may be filled with several coats of lacquer. Once the lacquer has dried, the defect should blend in with the fabric surface. Detects larger than 3/4 inch should be referred repair station.

Typically, separated laminations are repairable when they occur in the outside lamination. However, the repair must be done by a certified propeller repair station or the manufacturer. Additional repairs that can be made by a propeller repair station and a propeller manufacturer include large cracks that require an inlay and restoration of elongated bolt holes with metal inserts. In addition, broken sections may be repairable, depending on the location and severity of the break. However, the determination of reparability and actual repairs must be done by a repair station or the manufacturer.

When repair work is done to a wooden propeller blade, a protective coating must be reapplied to the wood. However, restoration of the protective coating could change the propeller blade's balance. Therefore, the propeller's balance must be checked after the blade has been refinished. If an out-of-balance condition exists, it may be necessary to apply more protective coating on one blade than the other to achieve final balancing.

As with all propellers, there are some defects that cannot be repaired. For wooden propellers these include:

1.   Crack or deep cut across the grain

2.   A split blade

3.   Separated laminations, except for the outside laminations of a fixed-pitch propeller

4.   Empty screw or rivet holes

5. Any appreciable warp

6. An appreciable portion of wood missing

7. An oversized crankshaft bore in a fixed-pitch propeller

8. Cracks between the crankshaft bore hole and bolt holes

9. Cracked internal laminations

10. Oversize or excessively elongated bolt holes

STORAGE

When a wood propeller is stored, it should be placed in a horizontal position to keep the moisture evenly distributed throughout the wood. In addition, the storage area should be cool, dark, dry, and well ventilated. Do not wrap the propeller in any material that seals it from the surrounding airflow. The reason for this is an airtight wrapping around wood propellers promotes wood decay.

The properties of aluminum alloys make aluminum propellers durable and relatively inexpensive to maintain. However, some types of damage can be severe enough to cause blade failure. Therefore, aluminum propellers must be carefully inspected at regular intervals. In addition, if any damage is discovered that jeopardizes the integrity of a propeller, it must be repaired before further flight.

BLADE INSPECTION

A requirement for both annual and 100-hour inspections includes checking for cracks, nicks; and properly. torqued or safetied bolts. As with a wood propeller, most inspections of a fixed-pitch aluminium propeller are conducted without removal. However, if operational problems such as vibrations occur, it may be necessary to remove the propeller for a detailed inspection of the hub area.

Removal procedures for a fixed-pitch aluminium propeller are the same as for a fixed-pitch wood propeller. However, the aluminium propeller may be heavier than a comparable sized wood propeller. If so, obtain help when necessary to support the propeller during removal and prevent damage to the propeller or personal injury.

Prior to an inspection, an aluminium propeller should be cleaned with a solution of mild soap and water using a soft brush or cloth to remove all dirt and grease. Acid or caustic cleaning materials should never be used on aluminium propellers because their use could lead to corrosion. Furthermore, avoid the use of power buffers, steel wool, steel brushes; or any other abrasive that may scratch or mar the blades. If a propeller has been subjected to salt water, it should be flushed with fresh water until all traces of salt have been removed. This should be accomplished as soon as possible after exposure to salt spray.

Once clean, aluminium blades are inspected for pitting, nicks, dents, cracks, and corrosion. Areas that are especially susceptible to damage include the leading edges and the blade face. To aid in the inspection process, you should inspect the entire propeller with a four-power magnifying glass. In addition, if a crack is suspected, a dye penetrant inspection should be performed. In many cases, a dye penetrant inspection will show whether visible lines and other marks are actually cracks or only scratches, saving the time and expense of unnecessary repairs.

Inspect the hub boss for damage and corrosion inside the centre bore and on the surfaces which mount on the crankshaft. Also, inspect the bolt holes for cracks, excessive wear, and proper dimensions. Light corrosion can be cleaned from the hub boss with sandpaper. The affected area may then be painted or treated to help prevent further corrosion. Propellers with damage dimensional wear, or heavy corrosion in the boss area should be referred to a repair station for appropriate repairs.

Repairs

When an inspection reveals surface damage such as nicks, scratches, or gouges on an aluminium alloy propeller blade, repairs should be made as soon as possible. By making prompt repairs, it will help eliminate stress concentration points which, in turn, helps to prevent cracks and fatigue failure. Defects on a bladed leading and trailing edges may be dressed out by using a combination of round and half round files. When a repair is complete, it should blend in smoothly with the edge and should not leave any sharp edges or angles. In all cases, the repair of surface detects on aluminium propeller blades must be made parallel to the length of the blade. In addition, the approximate maximum allowable size of a typical repair on a propeller edge is 1/ 8 inch deep by no more than 1 1/ 2 inches in length.

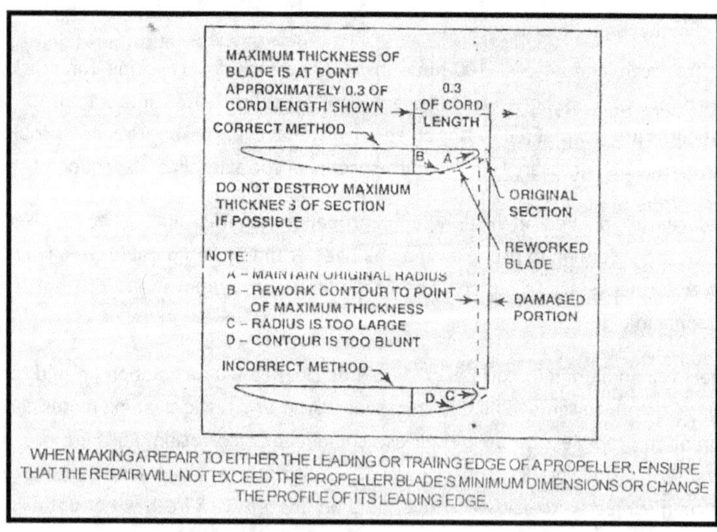

Repairs to the face and back of a blade are performed with a spoon-like riffle file which is used to dish out the damaged area. The maximum allowable repair size of a typical surface defect on a blade face or back is 1/16 inch deep by 3/8 inch wide by 1 inch long. In addition, all repairs must be finished by polishing with very fine sandpaper, moving the paper in a direction parallel to the length of the blade. Once sanded, the surface should be treated with Alodine, paint or other approved protective coating.

Damage in the shank area of a propeller blade cannot be repaired in the field and should be referred to an overhaul facility for corrective action. Since all forces acting on the propeller are concentrated at the shank, any damage in this area is critical. Furthermore, transverse cracks of any size render an aluminium alloy blade unrepairable.

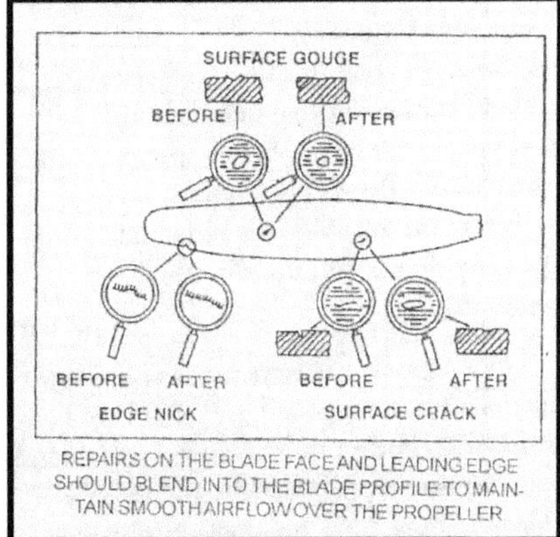

The bent blades can also be often be repaired. To determine if a blade is repairable begin by measuring the thickness of the blade where the bend is located. Once this is done, determine the blade station of the bend by measuring from the centre of the hub to the centre of the bend. With the centre of the bend located, mark the blade one inch on each side of the bend and place a protractor tangent to the one inch marks to determine the bend angle.

Many propeller manufacturers furnish charts that help a technician determine if a bend is repairable. In most cases, the chart consists of a graph with the blade station on one axis and the degree of bend on the opposite axis. When reading this type of chart, any bend below the graph line is repairable while any bend above the line is unrepairable.

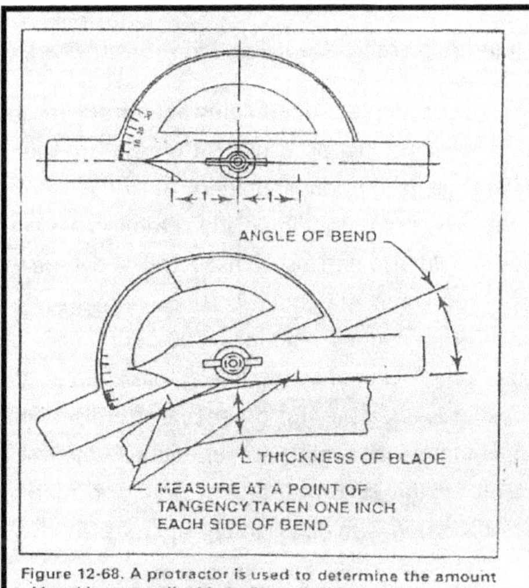

Figure 12-68. A protractor is used to determine the amount of bend in a propeller blade. Place the hinge over the center of the bend, set the protractor legs tangent to the blade one inch on each side of the bend centerline, and read the amount of bend in degrees on the protractor.

Once a repair has been made to an aluminium propeller, the propeller should be cleaned with an approved solvent. This helps remove all traces of dye penetrant materials used during an inspection and subsequent repair. If the propeller was painted, repaint the face of each blade with one coat of zinc chromate primer and two coats of flat black lacquer from the six inch station to the tip. The back of each blade should have the last four inches of the tip painted with one coat of zinc chromate primer and two coats of a high visibility colour.

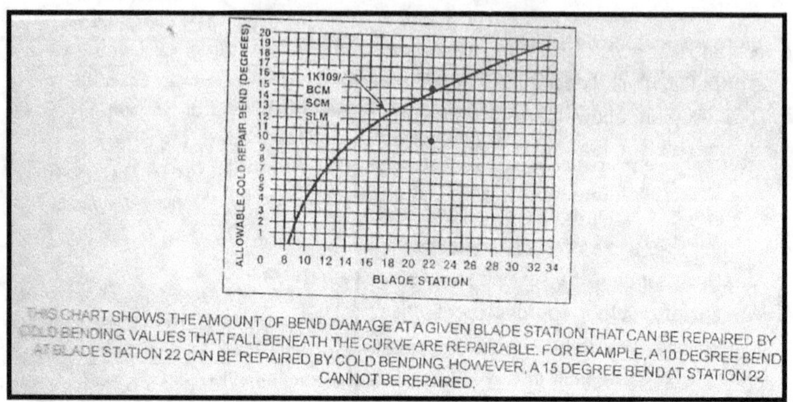

THIS CHART SHOWS THE AMOUNT OF BEND DAMAGE AT A GIVEN BLADE STATION THAT CAN BE REPAIRED BY COLD BENDING. VALUES THAT FALL BENEATH THE CURVE ARE REPAIRABLE. FOR EXAMPLE, A 10 DEGREE BEND AT BLADE STATION 22 CAN BE REPAIRED BY COLD BENDING. HOWEVER, A 15 DEGREE BEND AT STATION 22 CANNOT BE REPAIRED.

If a high polish is desired, a number of good grades of commercial metal polish are available. However, after completing the polishing operation, all traces of polish should be removed.

Ground-Adjustable Propellers

When inspecting a ground-adjustable propeller inspect the blades paying particular attention to the areas around the retention shoulders at the base of the blades. The corresponding blade retention areas of the hubs should also be closely inspected. A dye-penetrant inspection is recommended on the external surfaces in these areas during routine, 100-hour, and annual inspections: Unless complete disassembly of the propeller is necessary for other reasons, disassembly is not recommended at these inspection intervals. The reason for this is to reduce the wear and tear on the hub associated with disassembly and reassembly.

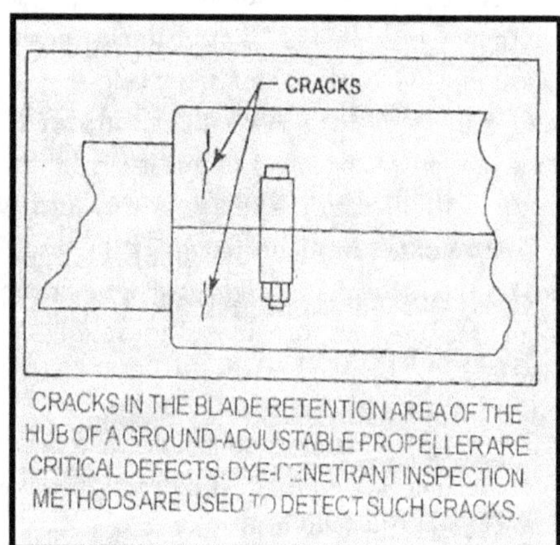

CRACKS IN THE BLADE RETENTION AREA OF THE HUB OF A GROUND-ADJUSTABLE PROPELLER ARE CRITICAL DEFECTS. DYE-PENETRANT INSPECTION METHODS ARE USED TO DETECT SUCH CRACKS.

Constant-speed propellers require a more detailed inspection and, in general, more maintenance than fixed-pitch propellers. For example, oil leaking from the propeller hub may' indicate a defective piston-to-cylinder a-ring. On some models, the ring can be replaced in the field by a technician following the procedures outlined in the propeller or aircraft service manual. On other models, the propeller must be returned to a propeller repair facility. Any seals, other than the piston-to-cylinder a ring which are found to be leaking require replacement by a propeller repair facility.

Hartzell Constant Speed Propellers

Hartzell constant-speed propeller systems require the same types of inspection, maintenance, and repair as other constant-speed systems. However, an additional inspection is recommended to check a steel hub for cracks. Magnetic particle inspection is the preferred method of inspection when checking steel propeller hubs for cracks.

If grease leakage is detected on a Hartzell propeller, determine the cause and correct it as soon as possible. The most common causes of grease leakage are loose, missing, or defective grease fittings, or zerks. Other causes could be loose blade clamps, defective blade clamp seals, and overlubrication of the blade-to-hub joints.

If a zerk fitting is, loose, missing or defective, it should be tightened or replaced as appropriate. Loose blade clamps should be torqued to the specified value for the particular model of propeller and resafetied. Check the blade angle to be certain that it does not change during retorquing.

Hartzell Feathering Compact Propellers

Inspection, maintenance, and repair procedures for a Hartzell feathering propeller system is the same as those for other Hartzell constant-speed systems. However, one additional check which should be accomplished at each 100-hour and annual inspection is the nitrogen charge within the propeller hub. If the charge is too low, it may not feather or respond properly to constant-speed operation and it may have a tendency. to overspend or surge. On the other hand, the propeller system may not reach full rpm and may feather upon engine shutdown if the nitrogen pressure is too high. When checking the nitrogen charge, ensure that the blades are latched in the low pitch position. If insufficient pressure exists, follow the manufacturer's instructions for servicing with nitrogen.

HAMILTON- STANDARD HYDROMATIC PROPELLERS

HydromatK propellers are inspected, maintained, and repaired in accordance with the same procedures as other constant-speed systems. Inspections primarily involve a check for proper operation, looking for oil leaks, and inspecting external oil lines for signs of deterioration or abrasion.

Oil leaks in the propeller are normally caused by a defective gasket or loose hardware. If oil covers all of the propeller, the likely cause is a leaking dome plug. Oil leakage: around the rear cone usually indicates a detective spider-shaft oil seal. If oil appears on the barrel immediately behind the dome, the dome gasket is leaking or the dome nut is loose_ The dome plug seal and the dome-to-barrel gasket can be replaced in the field.

Signs of leaking oil around the blade shank area or between the barrel halves could indicate loose hub bolts or defective gaskets. Loose hub bolts can be retorqued, but leaking gaskets must be replaced by an overhaul facility. The propeller is lubricated by engine operating oil, therefore it needs no other lubrication.

Propeller Lubrication

All the adjustable pitch propeller require inspections and servicing at regular intervals. Lubrication is, in many cases, one of the required servicing procedures. The grease used to lubricate a propeller must have the proper anti-friction and plasticity characteristics. In other words, an approved grease reduces the frictional resistant of moving parts and molds easily into any form under pressure.

Propeller -lubrication procedures are usually published in the manufacturer instructions along with oil and grease specifications. Experience indicates that water sometimes seeps into the propeller blade bearing assemblies of some propeller models. For this reason the propeller manufacturer's greasing schedule and recommended oil and grease specifications must be followed to ensure proper lubrication of moving parts.

Steel and Composite Propellers

Propellers made from steel or composite material may be cleaned and inspected in the same manner as wood or aluminium propellers. However, manufacturers of composite propellers may also include cleaning techniques and inspection items that are unique to composite materials. Manufacturer's instructions take precedence over general cleaning and inspection techniques.

With these propellers, you are restricted to inspections and cleaning. As you recall, any repairs made to correct defects in steel or composition propellers must be accomplished by an appropriately rated repair station or the manufacturer.

Blade Cuff Inspection

Some propeller blades are fitted with blade cuffs to improve airflow over the blade shank and cooling airflow through the engine. The blade cuffs must also be inspected and checked for proper clearance. Longitudinal clearance of constant-speed propeller blades or cuffs must be at least 1/2 inch between propeller parts and stationary parts of the aircraft. This clearance must be measured with the propeller blades feathered or in the most critical pitch configuration.

Governors

Inspection of governors is limited to checking for oil leaks and security of mounting. Maintenance consists of properly rigging the governor controls and verifying freedom of motion. Although you may remove and install propeller governors on an engine, inspections and repairs which require governor disassembly must be accomplished by a properly equipped and certified repair station.

Balancing

Exact propeller balance is critical to proper engine and propeller performance. Any time maintenance is conducted or a repair is made that adds or removes weight from a propeller, you must check the propeller's balance. For example, if a wood propeller is refinished, the new varnish can create an imbalance if it is unevenly applied. In another example, a metal blade that is shortened because of tip damage requires shortening of the opposite blade to maintain balance.

Propellers are balanced both statically or dynamically. A propeller is statically balanced when the propeller's centre of gravity coincides with its axis of rotation. On the other hand, a propeller is dynamically balanced when the centres of gravity of the blades rotate in the same plane of rotation.

Static Balancing

Static balancing is accomplished by using either the knife-edge method or the suspension method of the two static balancing methods, the knife-edge method is simpler and more accurate.

To balance a propeller using the knife-edge method, a test stand consisting of two hardened steel edges must be used. In addition, the test stand must be located in a room or area that is free from any air motion or heavy vibration.

Before you check a propeller's balance, you should first verify that the blade angles are all the same. If the blade angles are correct you can check a propeller's balance by following the listed sequence of operations

1.   Insert a bushing in the propeller hub bore hole.

2.   Insert a mandrel or arbor through the bushing to support the propeller on the balance knives.

3.   Place the propeller assembly so that the ends of the arbor are supported on the test stand. The propeller must be free to rotate.

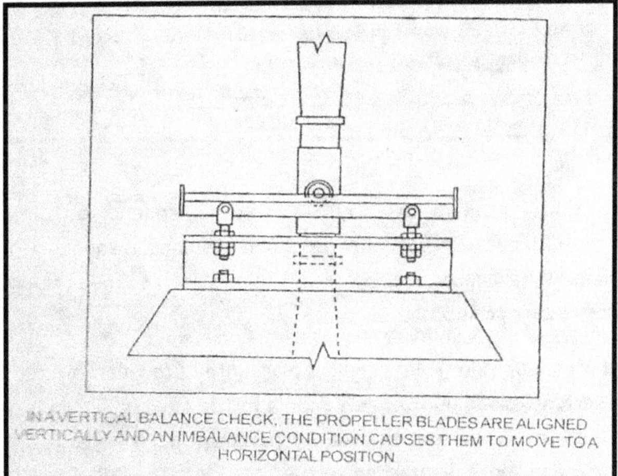

Once in the test stand, the propeller should be checked for horizontal and vertical balance. To check a two-bladed prOpeller assembly for vertical balance, position one blade in the vertical'-position. Next, repeat the vertical position check with the blade positions reversed from the first vertical check. If the propeller is balanced vertically it will remain in a vertical position regardless of which blade is pointing up. On the other hand, if a vertical imbalance exists, the propeller will have a tendency to come to rest in a horizontal position.

To check a two-bladed propeller assembly for horizontal balance, position the propeller in a horizontal position with both blades sticking straight out. If the propeller is horizontally balanced, it will remain in a horizontal position. On the other hand, if a horizontal imbalance exists, one blade will tend to move downward causing the propeller to come to rest in a vertical position.

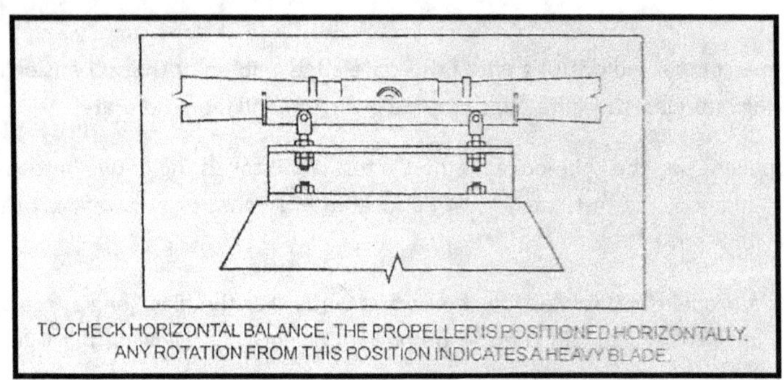

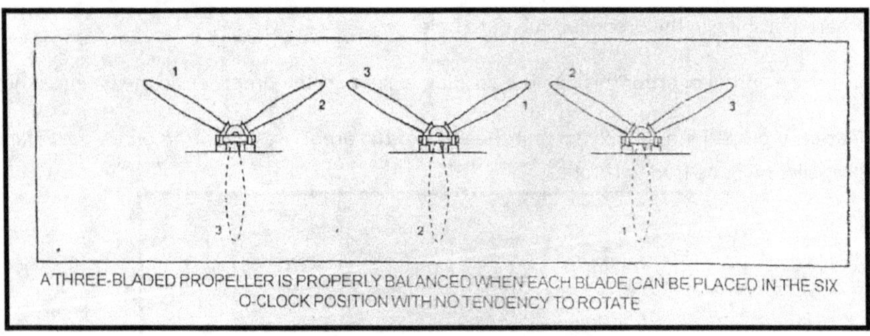

A two-bladed propeller that is properly balanced will have no tendency to rotate in any of the test positions. If the propeller balances perfectly in all described positions, it should also balance perfectly in all intermediate positions. When necessary, check for balance in intermediate positions to verify the check in the originally described positions.

Static balancing of a three-bladed propeller requires placing the propeller in three basic test positions. First, rotate the propeller until blade number one is pointing downward. Similarly, place blade number. In the suspension method for checking static balance the propeller is hung by a cord. A disk is firmly attached to the cord and a cylinder is attached to the propeller. Any imbalance is determined by the eccentricity between the disk and the cylinder.

Out-of-Balance Repairs

When a propeller assembly exhibits a definite tendency to rotate, certain corrections to remove the imbalance are allowed. The addition of permanent fixed weights is permitted at acceptable locations when the total weight of the propeller assembly is under allowable limits. Likewise, the removal of weight is permitted from acceptable locations when the total weight of the propeller assembly is equal to the allowable limit.

The location for removal or addition of weight on a propeller is determined by the propeller manufacturer. The method and point of application of balance corrections must be in accordance with the manufacturer's instructions. Typically, vertical imbalance is corrected by adding a metal weight on the light side of the hub 90 degrees from the propeller's horizontal centerline. On a wooden propeller, horizontal imbalance is corrected by adding or removing solder at the propeller blade tips. Horizontal balance correction on an aluminium propeller often involves the removal of small amounts of metal by filing.

Dynamic Balance

A propeller exhibits dynamic balance when the centres of gravity of similar propeller elements, such as the propeller blades, rotate in the same plane of rotation. A dynamic imbalance resulting from improper mass distribution is usually negligible if the blades on a propeller track within limits. One reason for this is that the length of the propeller assembly along the engine crankshaft is very short compared to its diameter. Another reason is the fact that the blades track the same plane perpendicular to the axis of the crankshaft.

Modern methods of checking dynamic balance require the propeller, spinner, and related equipment to be installed on the aircraft. With the engine running, electronic equipment senses and pinpoints the location of an imbalance. In addition, the test equipment typically determines the amount of weight required to correct the condition.

Checking Blade Angle

At times, it may be required to check the blade angle at a specific blade station. To do this, a universal propeller protractor is typically used.

The frame of a typical protractor is made of aluminium alloy with three square sides at 90 degree angles. A bubble spirit level mounted on one corner of the front of the frame swings out to indicate when the protractor is level. A movable ring is located inside the frame and is used to set the zero reference angle for blade angle measurements. The ring is engraved with Vernier index marks, which allow readings as small as one tenth of a degree. A centre disk is engraved with a degree scale from zero to 180 degrees, both positive and negative. The centre disk contains a spirit level to indicate when the centre disk is level.

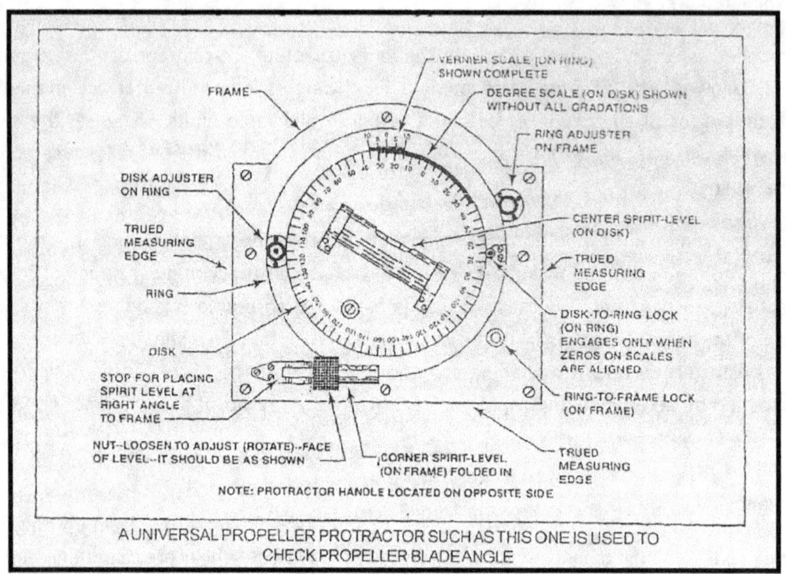

A UNIVERSAL PROPELLER PROTRACTOR SUCH AS THIS ONE IS USED TO
CHECK PROPELLER BLADE ANGLE

Before measuring a propeller blade angle, determine the reference blade station from the aircraft manufacturer's maintenance manual. Mark this reference station on the blade with chalk or with a grease pencil. Next, establish the reference plane from the engine crankshaft centerline. Do not reference the airframe attitude because some engines are installed at an angle to help counter the effects of torque. To zero the protractor, loosen the ring-to-frame lock, align the zeros on the disk and the ring, and engage the disk-to-ring lock. Place the edge of the protractor on a flat surface of the propeller hub that is either parallel to, or perpendicular to, the crankshaft centerline. Now, turn the ring adjuster until the spirit level in the center of the disk is level. The corner level should also be leveled. Now, tighten the ring-to-frame lock, and release the disk-to ring lock. The protractor is now aligned with the engine crankshaft.

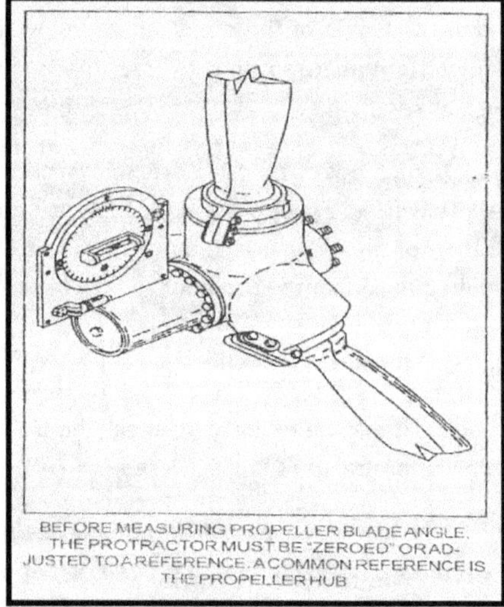

BEFORE MEASURING PROPELLER BLADE ANGLE,
THE PROTRACTOR MUST BE "ZEROED" OR AD-
JUSTED TO A REFERENCE. A COMMON REFERENCE IS
THE PROPELLER HUB

Once the protractor is zeroed, rotate the propeller until one blade is horizontal and place the protractor on the blade face at other reference station mark. Stand the same side of the propeller facing in the same direction you were when zeroing the protractor. If you desire to measure from the other direction, you must zero the protractor from that side. With the protractor resting on the face of the blade, turn the disk adjuster until the spirit level centers. Now read the blade angle using the zero line on the ring as the index. Accuracy in tenths of degrees can be read from the vernier scale. To measure the angle of another blade, rotate the desired blade to the same horizontal position and repeat the process.

If the face of the propeller blade is curved, use masking tape to attach a piece of 1/8 inch drill rod 1/2 inch from the leading and trailing edges. Once the rods are secure, measure the angle with the protractor resting on the rods.

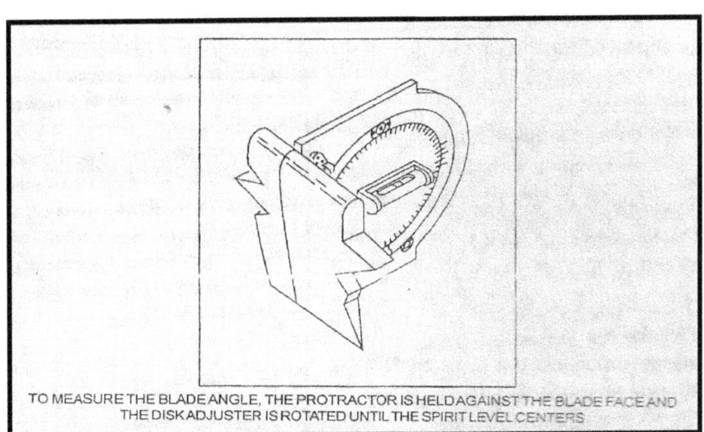

TO MEASURE THE BLADE ANGLE, THE PROTRACTOR IS HELD AGAINST THE BLADE FACE AND THE DISK ADJUSTER IS ROTATED UNTIL THE SPIRIT LEVEL CENTERS

If dissimilar blade angles exist on an aluminum fixed-pitch propeller, the blades can be repitched by a propeller repair station or the manufacturer. Consult the propeller repair facility and provide details on the amount of allowable pitch variation between blades.

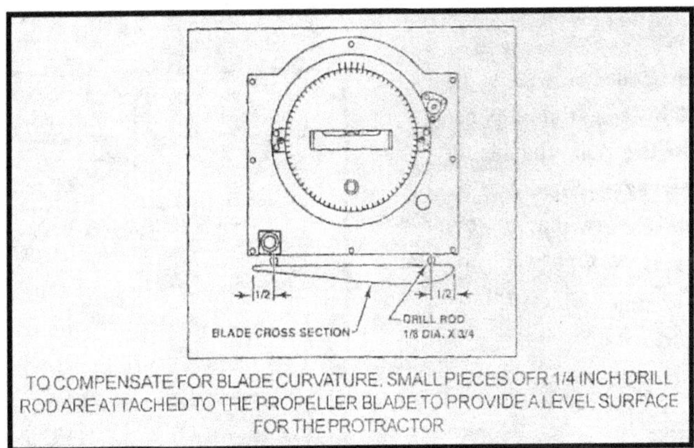

TO COMPENSATE FOR BLADE CURVATURE, SMALL PIECES OF R 1/4 INCH DRILL ROD ARE ATTACHED TO THE PROPELLER BLADE TO PROVIDE A LEVEL SURFACE FOR THE PROTRACTOR

Blade Angle Adjustments

All propeller systems other than fixed-pitch propellers require occasional blade angle adjustments. For example, ground-adjustable propellers are set to one blade angle while controllable pitch propellers require the setting of low and high blade angle limits. The method used to make blade angle adjustments depends on the propeller type. The following examples represent the more commonly used methods for adjusting blade angle.

Ground-Adjustable Propeller

To adjust the propeller blade angle on a ground-adjustable propeller, it must first determine the reference blade station that must be used. This information is typically contained in the propeller or aircraft maintenance manual. Once the reference station is known, check the specifications for the blade angle range approved for the aircraft. A typical range for a ground-adjustable propeller is from 7 to 15 degrees.

In most cases, a blade angle adjustment to a ground-adjustable propeller can be made with the propeller on the aircraft or on a propeller bench. To make the actual adjustment, begin by placing a grease pencil mark across the hub and blade to mark their relative positions. The mark provides visual identification of the original blade angle setting and a reference mark for adjustment to the new blade angle. Once the blades are marked, loosen the hub bolts or clamps and rotate the propeller to a horizontal position.

To change the blade angle, the hub halves must be separated slightly once the clamps or bolts are loosened. The blades may then be rotated in the hub until the desired blade angle is set. To help you rotate a given propeller blade, a propeller blade paddle is typically used if a blade binds, jiggle the blade as it is being rotated to the new angle.

Using the universal propeller protractor, check the blade angle after tightening the hub clamps or bolts. While loose in the hub, the blades droop slightly and will move a small amount as the hub is tightened. Because of this, you may be required to repeat the procedure a few times until the blades are set properly. Typically, an acceptable blade angle tolerance between the desired angle and the actual angle is 0.1 degrees. Once the tolerance is met, the propeller hardware can be properly torqued and safetied.

Counterweight

On a counterweighted propeller, the propeller blade angles are adjusted by means of a set of stop nuts on an index pin located tinder each counter-

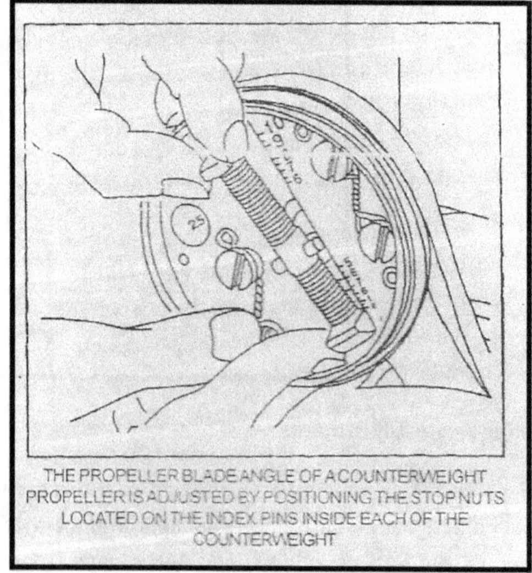

weight cap. To gain access to the index pin, remove the clevis pin which safeties the counterweight cap and remove the cap. Pull the index pin out of its recess in the counterweight or push it out from behind the counterweight bracket with a small tool.

Alongside the recess which holds the index pin is a scale calibrated with half degree marks and a numerical scale from zero to ten. This scale is used to adjust and set the stop nuts on the index pin.

The propeller blade index number, also known as the base setting, should be stamped in a lead plug located near the index pin recess. This number indicates the maximum blade angle for which the propeller was adjusted during its last overhaul. The maximum blade angle is typically 25 degrees and is used to calculate where the stop nuts on the index pin should be positioned. For example, if the blade index is 25 degrees and the aircraft specifications specify a low blade angle of 17 and a high angle of 22 degrees, the stop nut positions are determined by subtracting the appropriate blade angle from the blade index. Therefore, to set the 17-degree low blade angle, the stop nut is positioned on the index pin so that the edge toward the centre of the pin will align with the 8 degree (25 - 17 = 8) mark on the scale. On the other hand, to set the 22-degree high blade angle. the stop nut is positioned to line up its edge with the 3 degree (25 - 22 = 3) mark.

Once the stop nuts are set, the index pins are installed in the counterweights and the caps are replaced. With everything secured, the blades should be moved through their full range of travel. Once this is done, position the blades in their high blade angle and measure the blade angle at the spec-ified reference station. A common reference station for a counterweighted propeller is the 42-inch station. Next, move the blades to their low blade angle stop and check these angles. Make small adjustments to the stop nut positions as necessary to bring the angles within acceptable limits.

Hartzell Constant Speed Propellers

Hartzell steel hub propellers can be adjusted for the desired low blade angle by loosening the hub clamps and rotating the blades. However, you should realize that, anytime the low blade angle is changed, the high blade angle will also change. The reason for this is because the piston within the propeller hub can only travel a fixed amount. Once the desired blade angle is obtained, the clamps are re-torqued and safetied.

THE LOW PITCH SETTING FOR HARTZELL COMPACT PROPELLERS IS ADJUSTED WITH THE ADJUSTING SCREW ON THE HUB CYLINDER

The low pitch setting on a Hartzell compact propeller is adjusted with the adjusting screw on the hub cylinder. To make a blade adjustment, begin by loosening the jam nut on the adjusting screw and rotating the screw clockwise to increase the low blade angle, or counter-clockwise to decrease the angle. When the desired angle is set, retighten the jam nut. When changing the blade angles, always refer to the aircraft specifications and the propeller manufacturer's manual for instructions about specific propeller models.

Blade Tracking

Propeller blade tracking is a procedure which allows you to check the track of each propeller blade tip as it travels through its arc of rotation. In other words, by checking the tracking of a propeller, you compare the positions of the propeller blade tips relative to each other. This procedure is normally accomplished when troubleshooting a vibration problem or as a final check after balancing and reinstallating a propeller. Metal propellers up to six feet in diameter on light aircraft must track within 1/16 inch of each other. On the other hand, the track of a wood propeller should not be out more than 1/8 inch.

Before a propeller can be tracked, the aircraft must be locked in a stationary position. This is typically accomplished by chocking the wheels to prevent aircraft movement once this is done, place a fixed reference point on the ground that is within 1/4 inch of the propeller arc. This may be done by placing a board on blocks under the propeller arc and taping a piece of paper to the board.

A PROPELLER CAN BE TRACKED BY PLACING A BOARD WITHIN 1/4 INCH OF THE PROPELLER ARC. ROTATE THE PROPELLER AND MARK THE PATH EACH BLADE TIP FOLLOWS AS IT PASSES THE BOARD

With the reference point in place, rotate the propeller blade and mark the track of each blade. The maximum difference in track for all of the blades should not exceed the limits mentioned in the manual.

If the propeller track is off more than is allowed, the reason should be determined and the condition corrected. The easiest item to check is the torque of the propeller retaining bolts. If all bolts are properly torqued, the propeller should be removed to allow an inspection for the presence of debris or damage. In addition, it may be necessary to check the crank-shaft for alignment. If no problems are found, the excessive out-of-track condition may be corrected by placing shims between the inner flange and the propeller.

Troubleshooting

The origins of powerplant vibrations are sometimes difficult to pinpoint. To determine whether the vibrations are emanating from the engine or propeller, observe the propeller hub; dome, or spinner. With the engine running between 1,200 to 1,500 rpm, observe the hub or spinner for rotation on an absolutely- horizontal plane. If the propeller hub appears to swing in a slight orbit, or if the vibration becomes more apparent at higher rpms, the vibration is normally caused by the propeller. If the propeller hub oscillates, the difficulty is probably caused by engine vibration.

If excessive powerplant vibration is traced to the propeller, the problem could be one of several things. For example, dissimilar blade angle settings can lead to an uneven thrust distribution between propeller blades which. in turn can lead to vibration. Additional causes of propeller vibration include propeller blade imbalance, improper blade tracking, a loose retaining nut, loose hub hardware, or excessive crankshaft spline r" wear.

In addition to vibration problems, a malfunctioning pitch-changing mechanism may require troubleshooting. For example, sludge in oil passages or the pitch selector valve of a two-position propeller may cause slow or erratic responses to pitch-change commands. Furthermore, erratic or jerky blade movement during pitch changes may be caused by trapped air in the cylinder. Cycling the propeller through its pitch-change operation several times usually purges air from the system. In addition, the linkage from the cockpit pitch control lever to the selector valve may become loose and fail to activate the selector valve. Operations in a marine environment can cause salt-water corrosion around the propeller cylinder and piston, leading to sporadic and unreliable pitch changes.

Hamilton Standard Hydromatic

Troubleshooting procedures and solutions discussed for other systems are generally applicable to the feathering hydromatic system. If the propeller fails to respond to the cockpit propeller control lever, but can be feathered and unfeathered, the cause is most likely a failure of the governor or governor control system. If the propeller fails to feather, check the system for electrical faults or for open wiring to the electrical components.

If the propeller fails to unfeather after feathering normally, the distributor valve is not shifting. On the other hand, if the propeller feathers and immediately unfeathers, the problem may be a short circuit in the holding coil wiring or an open circuit in the pressure cut out switch or its associated wiring. The same problem occurs if the feather button is short-circuited internally.

Sluggish movement of the propeller may be the result of a buildup of sludge in the propeller dome or f worn out piston-to-dome seal inside the dome. Sticking cam rollers may also interfere with smooth pitch-change movement and feathering operations.

Erratic or jerky operation of the propeller is an indication of the wrong preload shim being used between the dome and barrel assemblies. If this is the case, the dome will have to be removed so the proper shim can be installed.

Propeller Installation

The method used to attach a propeller to an engine crankshaft varies with the design of the crankshaft. Currently, there are three types of crankshafts used on aircraft engines, the flanged crankshaft, the tapered crankshaft, and the splined crankshaft. The general installation, procedures for all three types are discussed in the following paragraphs. For specific instructions, it should refer to the aircraft and engine maintenance manuals.

Flanged Shaft

Flanged propeller shafts are used on most Horizontally opposed reciprocating engines and some turboprop engines. The front of the crankshaft is formed into a flange four to eight inches across, perpendicular to the crankshaft centerline. Mounting bolt holes and dowel pin holes are machined into the flange. Some flanges have threaded inserts pressed into the bolt holes.

Before installing a propeller on a flanged shaft, inspect the flange for corrosion, nicks, burrs, and other surface defects. In addition, the bolt holes and threaded inserts should be clean and in good condition. Any defects found should be repaired in accordance with the engine manufacturer's recommendations. Light corrosion can typically be removed with very fine sandpaper; however, if a bent flange is suspected, a run out inspection should be 1".94 performed. If you do have to remove corrosion, clean the flange after sanding and check for smoothness. Once this is done, apply a light coat of engine oil or anti-seize compound to the flange for corrosion prevention and ease of future propeller removal.

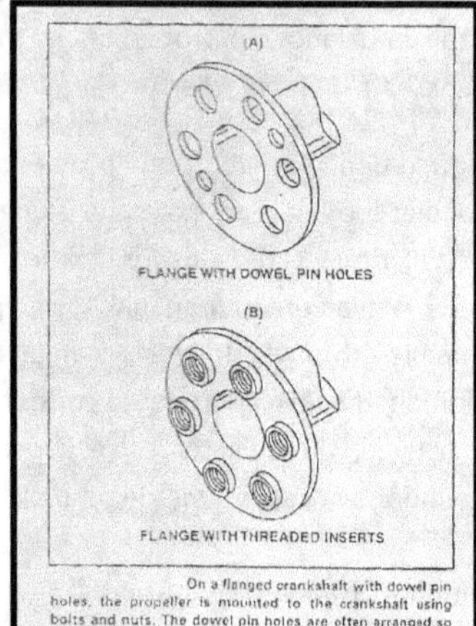

On a flanged crankshaft with dowel pin holes, the propeller is mounted to the crankshaft using bolts and nuts. The dowel pin holes are often arranged so the propeller can mount in only one position. (B) — Most installations utilize threaded inserts which are pressed into the crankshaft to eliminate the use of nuts.

Fixed-Pitch Propellers

Before installing a fixed-pitch propeller on a flanged shaft, inspect the mounting surface of the propeller to verify that it is clean and smooth. The attaching bolts should be in good condition and inspected for cracks with either a dye penetrant or magnetic particle inspection process. Washers and nuts should also be inspected, and new fiber lock nuts used if they are required in the installation.

Most flanges that use dowel pins allow the propeller to mount on the shaft in only one position. If there is no dowel, install the propeller in the position specified by the aircraft or engine maintenance manual. This is important because propeller position is critical for maximum engine life in some installations. If no position is specified on a four cylinder horizontally opposed engine, the propeller should be installed with the blades at the 10 o'clock and 4 o'clock positions when the engine is stopped. This reduces vibration in many instances and puts the propeller in the lest position for 1-land propping.

When installing a propeller on a four-cylinder opposed engine, one of the blades should come to rest at the ten o'clock position to help reduce vibration and facilitate hand propping.

After attaching the bolts, washers, and nuts, tighten all of the bolts finger-tight. Then, use an approved torque wrench to tighten the bolts to a specified value in the recommended sequence. A typical torque value is 35 foot-pounds or higher for metal propellers and approximately 25 foot-pounds for wood propellers. In addition, a typical sequence requires you to torque the bolts in a crossing pattern.

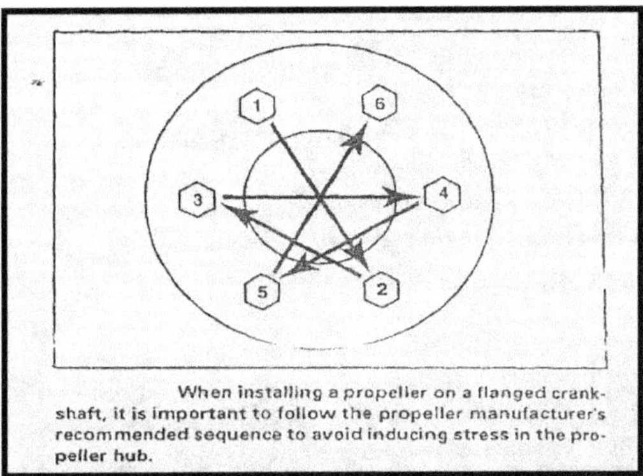

When installing a propeller on a flanged crankshaft, it is important to follow the propeller manufacturer's recommended sequence to avoid inducing stress in the propeller hub.

When a skull cap spinner is used, a mounting bracket is installed behind two of the propeller mounting bolts. Once the mounting bracket is installed, the skull cap is attached to the bracket with a bolt and washer.

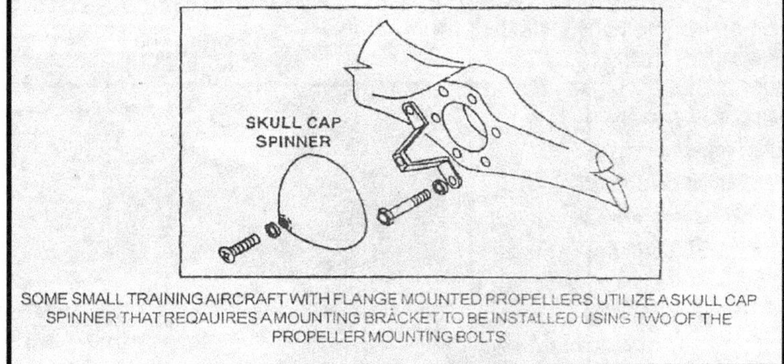

SOME SMALL TRAINING AIRCRAFT WITH FLANGE MOUNTED PROPELLERS UTILIZE A SKULL CAP SPINNER THAT REQAUIRES A MOUNTING BRACKET TO BE INSTALLED USING TWO OF THE PROPELLER MOUNTING BOLTS

If a full spinner is used, a rear bulkhead is slipped on the flange before the propeller is installed. After mounting the propeller, a front bulkhead is placed on the front of the hub boss before the bolts are inserted. After the bolts are tightened and safetied, the spinner is installed with machine screws. The machine screws are inserted through the spinner into nut plates on the bulkheads. If the spinner is indexed, line up the index marks during installation to avoid vibration.

Constant-Speed Propellers

Some Hartzell steel hub propellers and all Hartzell compact propellers are designed to mount on flanged crankshafts. However, before you mount a constant-speed propeller on the crankshaft, you should lubricate the 0-ring in the rear of the hub with a light coat of engine oil. Once this is done, you can carefully mount the propeller on the flange. When doing this, pay particular attention to the 0-ring to keep it from being damaged.

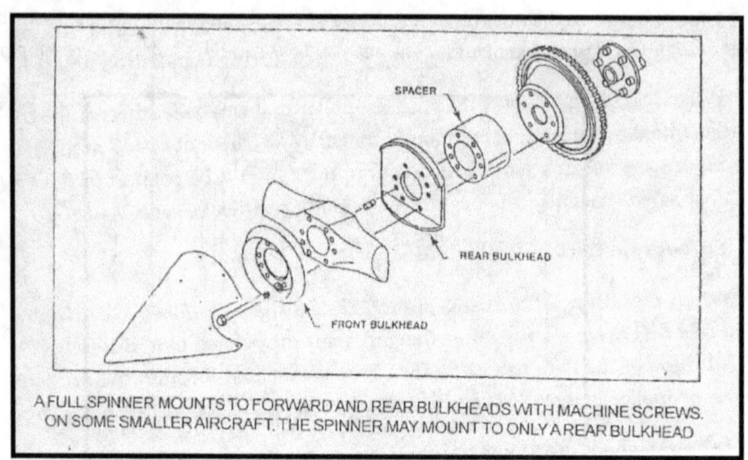

A FULL SPINNER MOUNTS TO FORWARD AND REAR BULKHEADS WITH MACHINE SCREWS. ON SOME SMALLER AIRCRAFT, THE SPINNER MAY MOUNT TO ONLY A REAR BULKHEAD

McCauley constant-speed, propellers are also installed on flanged crankshafts. Like Hartzell propellers, an 0-ring in the rear of the hub must be lubricated with a light coat of engine oil to allow its movement as the propeller is secured to the crankshaft flange. A dry 0-ring can tear and become pinched during installation unless lubricated.

When installing a constant-speed propeller that can be feathered, the installation and adjustment procedures are similar to that of other constant-speed models. However, if the blades are left in a feathered position, they should be rotated to their low pitch angle. For safety reasons, it is best to use a blade paddle on each blade.

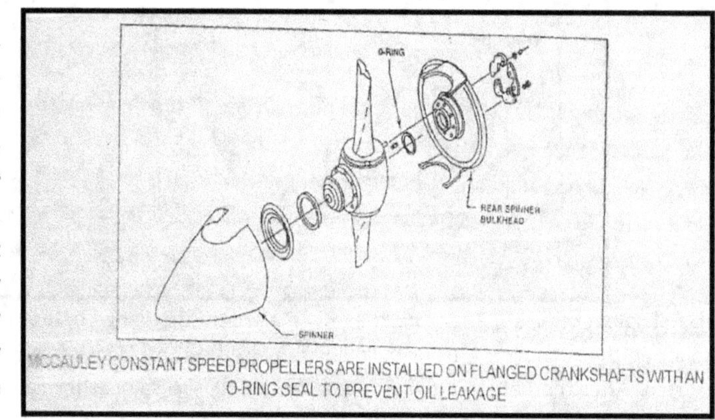

McCAULEY CONSTANT SPEED PROPELLERS ARE INSTALLED ON FLANGED CRANKSHAFTS WITH AN O-RING SEAL TO PREVENT OIL LEAKAGE

Turbopropellers

When installing a constant-speed, reversing propeller, use the same basic procedures that are used for other flanged shaft propellers. One difference, however, is the addition of the Beta tube. The Beta tube is installed through the propeller piston after the propeller is installed, and is bolted to the forward part of the piston.

Tapered shaft crankshafts are found on older engines that produce low horsepower. This type of crankshaft requires a hub to adapt the propeller to the shaft. To prevent the propeller from rotating on the shaft, a large keyway is cut into the crankshaft taper and the propeller so that a key can hold the propeller in place.

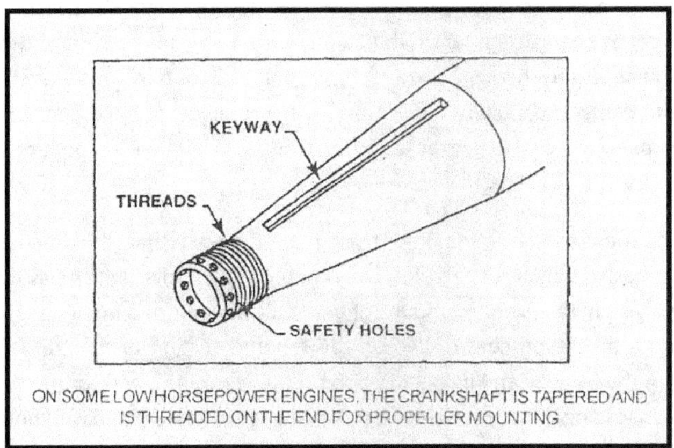

ON SOME LOW HORSEPOWER ENGINES, THE CRANKSHAFT IS TAPERED AND IS THREADED ON THE END FOR PROPELLER MOUNTING

When installing a wood propeller on a tapered shaft, the propeller boss is installed over the adapter hub and a faceplate is placed between the boss and mounting bolts. This faceplate distributes the compression load of the bolts over the entire surface of the boss. If a new fixed-pitch wood propeller is installed, inspect the mounting bolts for tightness after the first flight and again after the first 25 flight hours.

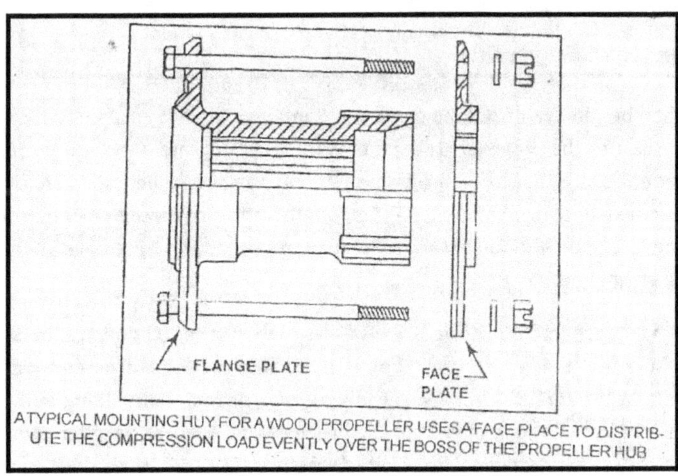

A TYPICAL MOUNTING HUY FOR A WOOD PROPELLER USES A FACE PLACE TO DISTRIBUTE THE COMPRESSION LOAD EVENTLY OVER THE BOSS OF THE PROPELLER HUB

Before installing the propeller on the crankshaft, the shaft must be carefully inspected for corrosion, thread condition, cracks, and wear in the keyway area. If cracks are allowed to develop in the keyway, they can spread rapidly and eventually cause crank-shaft failure. It is good practice to inspect the key-way with dye penetrant at every 100-hour or annual inspection. Any minor surface defects found during the pre-installation inspection should be dressed or polished out in accordance with the engine manufacturer's maintenance manual. In addition, the propeller hub components and mounting hardware should be inspected for wear, cracks, and corrosion. Defective components must be replaced or repaired as necessary.

permanently installing the propeller, a trial fit of the hub on the crankshaft should be done using a liquid transfer ink such as Prussian Blue. Prussian Blue is a dark blue ink, or dye, which has the consistency of a light grease. This dye visibly reveals the amount of contact between two mating surfaces. To do a trial fit, begin by applying a thin, even coat of dye on the tapered section of the crank-shaft. Once this is done, place the key in the keyway and install the hub on the crankshaft and torque the retaining nut. In practice, the hub, snap ring, and retaining nut are never disassembled. If, however, they were disassembled for inspection or repair, place the retaining nut against the hub and install the puller snap ring. Once assembled, the retaining nut may be torqued.

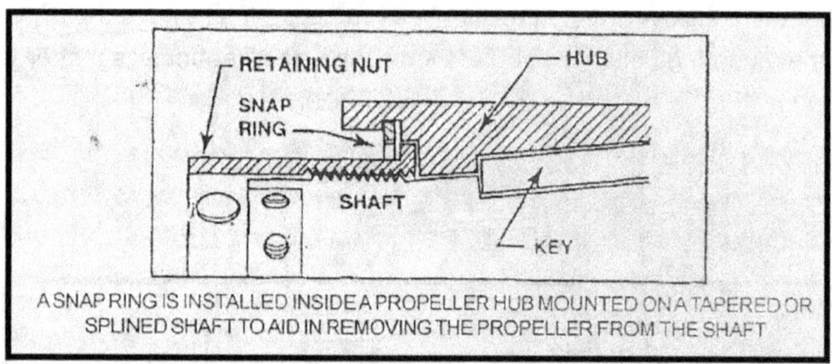

A SNAP RING IS INSTALLED INSIDE A PROPELLER HUB MOUNTED ON A TAPERED OR SPLINED SHAFT TO AID IN REMOVING THE PROPELLER FROM THE SHAFT

The hub should then be removed from the crank-shaft and inspected for the amount of ink transferred from the tapered shaft to the propeller. The ink transfer must indicate a minimum contact area of 70 percent. If -insufficient contact is found, the crankshaft and hub should be inspected for the cause. The mating surfaces can be lapped with a polishing compound -until. a minimum of 70 percent contact area is achieved. After this is done, thoroughly clean the hub and crankshaft to remove all traces of Prussian Blue and polishing compound.

Once the minimum contact area is achieved, apply a very light coat of oil or antiseize compound to the crankshaft. Make sure that the key is installed properly, then place the hub assembly and propeller on the shaft. Be sure that the threads on the shaft and nut are clean and dry, then verify that the puller snap ring is in place before torquing the nut to the proper value. Failure to tighten the retaining nut to the proper torque results in play between the propeller, front cone and rear cone. Any space between the cones and the propeller produces galling and wear on their surfaces. Safety the retaining nut to complete the installation.

Splined Shaft

Splined crankshafts are found on most radial engines, some horizontally opposed, and some inline engines. The splined shaft has grooves and splines of equal dimensions and a double width master spline to ensure that a hub will fit on the shaft in only one position.

Before installing a propeller on a splined shaft, inspect the crankshaft for cracks, surface defects, and corrosion: If any defects exist, repair them in accordance with the engine manufacturer's instructions. Crankshaft and hub splines are inspected for wear with a go/no-go gauge which is 0.002 inch larger than the maximum space allowed between the splines. The splines are serviceable if the gauge cannot be inserted between the splines for more than 20 percent of the spline length. If the gauge goes in more than 20 percent of the way, the hub or the crankshaft is unairworthy and must be replaced.

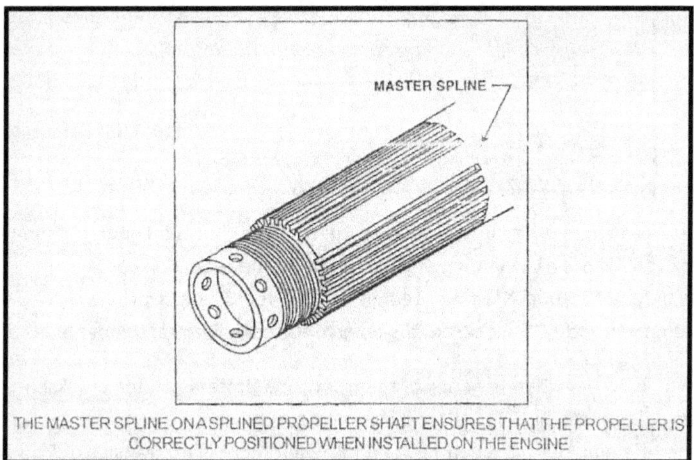

THE MASTER SPLINE ON A SPLINED PROPELLER SHAFT ENSURES THAT THE PROPELLER IS CORRECTLY POSITIONED WHEN INSTALLED ON THE ENGINE

To help ensure that the propeller hub is centered on the crankshaft, a front and rear cone are installed on each side of the propeller hub. The rear cone is typically made of bronze and is split to allow flexibility during installation and to ensure a tight fit The front cone, on the other hand, is made in two pieces as a matched set. The two halves are marked with a serial number to identify them as mates in a set.

In addition to the front and rear cones, a large retaining nut is used to tighten and hold the propeller in place. The retaining nut threads onto the end of the splined and presses against the front cone to sandwich the propeller tightly between the and rear cones.

Like the tapered shaft, a trial installation of the pro-pellet- should be completed a proper fit. To do the trial installation, begin by applying a thin coat of --- glue to the rear cone. Next, slip the rear cone and bronze spacer onto the pushing them all the way back on the shaft. With the rear cone in place, align hub on the master spline and push the hub back against the rear cone. Coat the front cone halves with Prussian Blue and place them around the lip of the retaining nut. Install the nut in the hub and tighten it to the proper torque.

After the retaining nut is torqued, immediately remove the retaining nut and front cone and note thd amount of Prussian Blue transferred to the hub. A minimum of 70 percent contact is required. Then, remove the hub from the crankshaft and note the transfer of dye from the rear cone. As with the front cone, a minimum of 70 percent contact is required. If contact is insufficient, lap the hub to the cones using special lapping tools and fixtures.

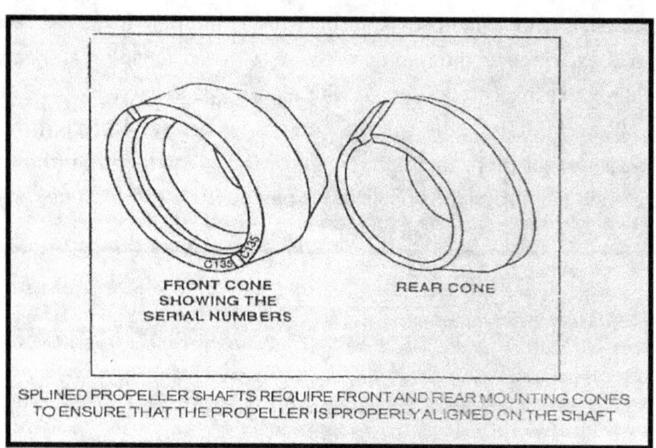

If no dye is transferred from the rear cone during the transfer check, a condition known as rear cone bottoming may exist. This occurs when the apex, or point, of the rear cone contacts the land on the rear seat of the hub before the hub becomes seated on the rear cone. One way to correct rear cone bottoming is to remove up to 1/16 inch from the apex of the cone with sandpaper on a surface plate.

Front cone bottoming occurs when the apex of the front cone bottoms on the crankshaft splines, before it has a chance to seat on the hub. Front cone bottoming is indicated by either the hub being loose on the shaft after the retaining nut has been torqued, or when there is no transfer of Prussian Blue to the front hub seat. Correct front cone bottoming by using a spacer of no more than 1/8 inch thickness behind the rear cone. This moves the hub forward, enabling the hub to properly seat on the front cone.

Once a proper fit between the hub and splined shaft is ensured, reinstall the rear cone and permanently mount the propeller on the shaft. As you recall, the position of the propeller on the hub in relation to the master spline is predetermined. Some installations require a certain blade to align with the master spline while other installations require the blades be perpendicular to the master spline position. Therefore, be sure to consult the engine maintenance manual for the requirements of a particular Ira installation.

Propeller Safetying

Once a propeller is properly torqued, it must be safetied. There is no one correct way to safety a propeller installation because of the many different types of installations. For this reason, the discussion of safetying methods is limited to the types more re' commonly used.

A flanged shaft installation has the largest variety of safety methods because of its many variations. If the flange has threaded inserts installed, the propeller is held on by bolts screwed into the inserts. In this case, the bolt heads are drilled and safetied with 0.04 inch stainless steel safety wire, using standard safety wire procedures.

If threaded inserts are not pressed into the flange, bolts and nuts are the propeller in place. Some installations use fiber lock nuts which require no safetying, a used to hold but the nuts should be replaced each time the propeller is removed. For installations using castellated nuts and drilled bolts, the nuts are safetied to the bolts with cotter pins.

The retaining nuts for tapered and splined shaft installations are safetied in the same way. In addition, a clevis pin is installed through the safety holes in the retaining nut and crankshaft. The clevis pin must be positioned with the head toward the center of the crankshaft. This allows centrifugal force to hold the clevis pin tightly in the hole against its flanged head.

An operational check should be conducted once a constant-speed propeller has been installed and safetied. To conduct this check, follow ground run-up procedures for the aircraft you are operating and position the aircraft for maximum safety. The first time a newly installed propeller operates at high rpm on an engine, it is always wise to be alert to the hazards of possible propeller failure.

All adjustable propeller systems share common features in regard to their control configuration. Propeller controls must be rigged so that an increase in rpm is obtained by moving the controls forward and a decrease in rpm is caused by moving the controls aft. Furthermore, engine throttles must be arranged so that forward thrust is increased by forward movement of the control, and decreased thrust is obtained by aft movement of the throttle.

When running-up an engine and testing a newly installed Hydromatic propeller, it is necessary to exercise the propeller several times. This is done by moving the governor control through its entire range of travel several times to free the dome of entrapped air.

Once all ground checks and adjustments are successfully completed, a test flight should be conducted. The test flight verifies the propeller system response to dynamic loads and determines if any other adjustments are necessary. After the test flight, check for oil leaks and component security.

FUEL NOZZLE TESTING

Testing PT6A Fuel Nozzles
Because the information in this section is not totally complete, the manufacturer's maintenance manual as well as appropriate safety precautions should always be used when performing any type of maintenance on fuel nozzles. An example of a fuel nozzle for a PT6A engine is shown in figure. It consists of a fuel manifold adapter (5), a sheath (3), a locking plate (6), a tabwasher (7), and a fuel nozzle (8). Due to the important nature of the spray pattern of a nozzle, it is necessary to clean and test fuel nozzles at certain intervals.

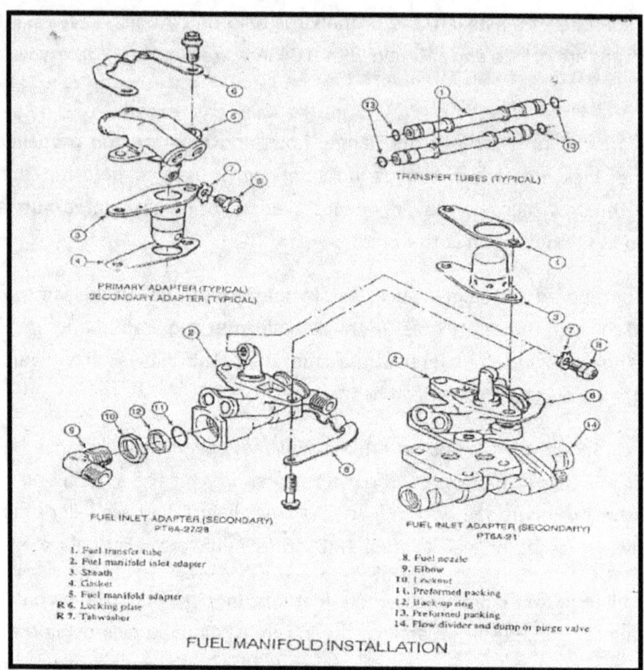

FUEL MANIFOLD INSTALLATION

The procedure for testing fuel nozzles begins with lubricating the nozzles with fuel, then installing the nozzles into the manifold adapters (5 or 2) using new tabwashers for each nozzle. Tighten the nozzle assemblies and torque, but do not bend the lugs of the tab-washers until testing is completed.

Fuel nozzles are subjected to two separate tests, the leakage test and the functional test. The leakage test is used to make sure the nozzle is not leaking at the adapter connection. The functional test is used to observe the spray pattern.

To perform the leakage test, back off the torque screw (6) and the setscrew (4), loosen the pivot screw (8), and rotate the pivot block. This will provide adequate space between the plugs in the pivot block and the plugs in the upright end of the fixture (2) to allow the installation of the adapter and nozzle assembly (3). Position the adapter end nozzle assembly (3) between the pivot block and the end of the fixture (2) with the nozzle facing toward the plastic pad. (5).

Insert the plugs of the pivot block into the ports in the adapter and, by rotating die block, insert the plugs of the fixture (2) into the opposite ports in the adapter. Do not dislodge the preformed packings from the seal grooves in the plugs during this with the plugs fully inserted into the adapter ports, hold the parts firmly - - -- the pivot screw (8).

Close off the nozzle orifice by turning the setscrew (4) until it just makes contact with the rear of the adapter behind the nozzle. Turn the torque screw (6) until the plastic pad (5) seats on the nozzle orifice. Tighten the setscrew (4) and torque the torque screw (6), simultaneously, fo ensure that the plastic pad (5) closes the nozzle orifice without distortion of the adapter. Tighten the locknut on the setscrew (4).

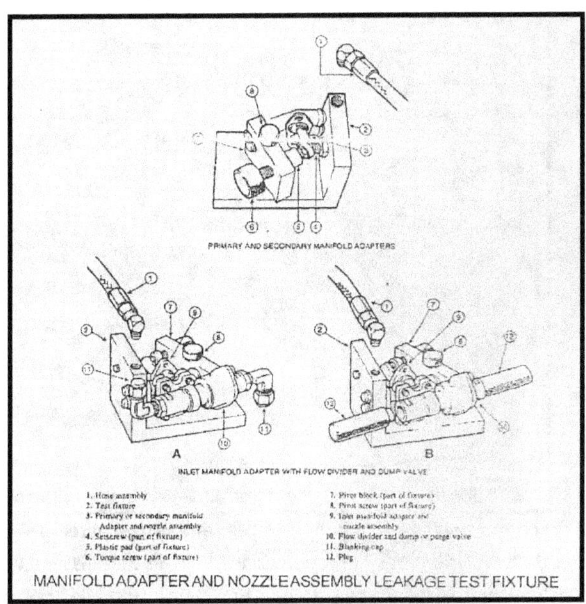

MANIFOLD ADAPTER AND NOZZLE ASSEMBLY LEAKAGE TEST FIXTURE

When testing the inlet manifold adapter and nozzle assembly (9) with the attached flow divider and dump or purge valve (10), close off the elbows on the flow divider and dump or purge valve (10) with caps (11). If elbows are not installed, close off ports in the flow divider and dump or purge valve (10) with plugs (12).

Check for leakage between the nozzle and the adapter by connecting the hose assembly (1) to a supply of clean, dry compressed air or nitrogen and applying a pressure of 500 psig to the test fixture (2). Check for leakage using leak check fluid or by immersing in a petroleum solvent. No leakage is permitted.

The functional test consists of flowing fluid through the nozzle and observing the spray pattern. There are several terms used in this text that describe specified test conditions. For example, the term "onion" describes a spray condition which sometimes occurs at low flow rates when the spray exhibits a distinct. shape. "Streakiness" is defined as variation in spray quantity among different parts of the spray cone and appears as a darker streak in the spray. It is specified as a percentage variation from nominal. "Spitting" is a condition which exists when large drops of unatomized fuel occur intermittently and usually on the outside of the spray cone. "Drooling" is a condition which occurs when large drops of unatomized fuel form on the nozzle face.

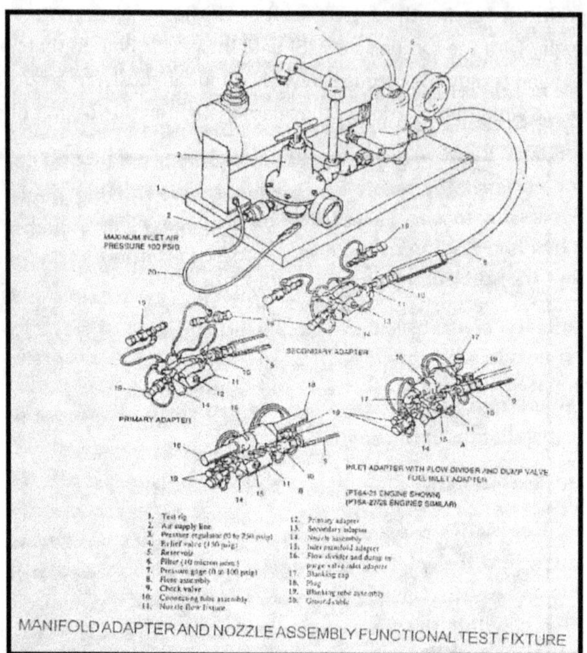

MANIFOLD ADAPTER AND NOZZLE ASSEMBLY FUNCTIONAL TEST FIXTURE

Figure illustrates a functional test of the primary adapter and nozzle assemblies. closing off the primary tube assembly using the blanking tubes (19) and connecting the tube assembly (10) of the flow fixture (11) into the ports as shown, the two remaining blanking tube assemblies (19) are not required for this test. Place the adapter (12) into the flow fixture (11) and ensure that the collars of the tube assemblies (19 and 10) are located on the inside of the fixture's flanges with the tubes in their appropriate slots. Secure the adapter to the fixture with the screws and nuts provided.

Connect the tube assembly (10) to the check valve (9) on the hose assembly (8), and tighten the coupling nut securely to prevent leakage. Connect an air supply line (2) with a maximum inlet pressure of 100 psig to the pres-sure regulator (3) on the test rig (1). Slowly adjust the pressure regulator (3) until 20.0 psig is indicated on the pressure gage (7). With the nozzle pointing downward, observe the spray pattern at the nozzle. An open spray must be observed, free from spitting and drooling. An onion spray may be evident between the nozzle tip and the open spray; this spray pattern, however, is not a requirement.

Adjust the pressure regulator (3) to increase the pressure to 60 psig as indicated on the gage (7). The volume of spray should increase and be evenly spread around the center axis of the nozzle orifice. If streakiness of more than 20 percent is evident, reject the nozzle. Spitting, drooling, and streakiness may be caused by external carbon deposits around nozzle orifices. Remove these deposits by lightly brushing the face of the nozzle with a bronze or non-metallic bristle brush while fuel is flowing through the orifice. Reduce the flow pressure to zero, as indicated on the gage (7), by adjusting the pressure regulator (3). When fuel flow from the nozzle stops, disconnect and close off the hose assembly (8) from the connecting tube assembly (10).

Upon completion of a satisfactory test, bend the lugs of each tab washer over the hexagon on each nozzle assembly.

Place the adapter and the nozzle assembly in a clean, dustproof container until it is required for installation in the engine.

Component Maintenance

Fan Blades. Fan blades receive damage from time to time because of foreign objects being drawn into the inlet of the engine. Small rocks cause nicks which are usually repairable as specified in the maintenance manual. Typically, a small nick may be repaired if it is within the dimensions specified. Figure is an example of repair' limits for the first-stage fan blade of a JT8D engine. The cuts made in the process of repairing the blade are termed "flyback cuts." If fan blade damage is such that all damaged sections can be removed within the limitations, the blade can be continued in service for a maximum of 20 h. The repair must adhere to any combination of limits shown for cuts 1, 2, 3, and 4; and the blade may be repaired up to the maximum dimension defined by the envelope created by all four cuts.

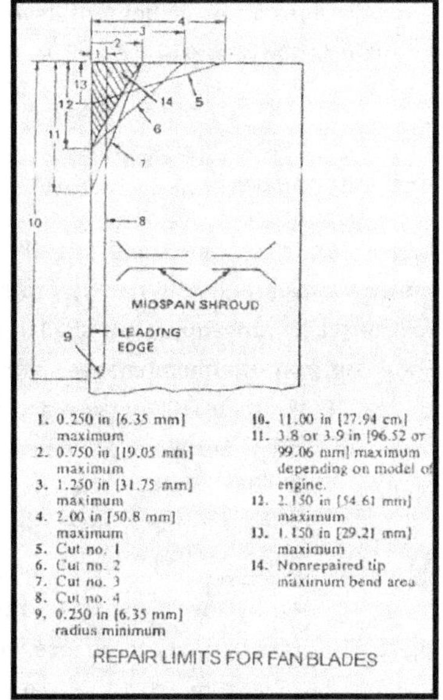

1. 0.250 in [6.35 mm] maximum
2. 0.750 in [19.05 mm] maximum
3. 1.250 in [31.75 mm] maximum
4. 2.00 in [50.8 mm] maximum
5. Cut no. 1
6. Cut no. 2
7. Cut no. 3
8. Cut no. 4
9. 0.250 in [6.35 mm] radius minimum
10. 11.00 in [27.94 cm]
11. 3.8 or 3.9 in [96.52 or 99.06 mm] maximum depending on model of engine.
12. 2.150 in [54.61 mm] maximum
13. 1.150 in [29.21 mm] maximum
14. Nonrepaired tip maximum bend area

REPAIR LIMITS FOR FAN BLADES

It will be noted that the leading edge of the blade can be cut back a distance of 0.250 in for a distance of 11 in along the blade. Toward the tip, deeper cuts can be made as shown.

For blades with FOD (foreign-object damage) confined to the blade tip only, repair may be made and the blade continued in service provided that the repair adheres to the limits shown for cut 1,2, or 3 and that the blade is repaired up to the maximum dimension of only one of the permissible cuts.

In the repair of fan blades, certain conditions are specified. For example, all repair cuts must have a length-to-depth ratio greater than 4: 1. Contours must be smooth and continuous, with a minimum radius of 0.250 in. The leading-edge contour after repair should conform as nearly as possible to the original. Repaired areas must be checked with a dye or fluorescent penetrant to ensure that there are no cracks.

The repair of fan blades while the fan rotor is installed in the engine requires that the area to be reworked be completely masked off to ensure that no metal splatter can strike any other blade or disk surface. Cutting is accomplished with a 2-in cutting wheel mounted in an air chuck operating at 18,000 rpm maximum. A minimum of 0.060 in of material must be left for hand filing and polishing to ensure removal of any heat-affected areas.

Shingled blades may be unshingled and continued in service for a maximum of h provided that they can be unshingled without further damage and that inspection shows that the midspan shroud of a shingled blade has not hit the airfoil section of an adjacent blade or the radius between the airfoil and the midspan shroud of the adjacent blade. Blades showing evidence of having been hit in this manner must be removed from service before further flight. After 20 h of service, shingled blades should be removed and subjected to overhaul-type inspection.

Compressor Blades

Compressor blades are subject to the same type of damage encountered by fan blades, and the repair procedures are similar. Figure is adapted from the maintenance manual for the Pratt & Whitney JT8D engine and shows some of the permissible repairs for compressor blades. Note that there are definite limits on the depth of a cut that is allowed in removing a nick, scratch, or other - damage caused by the ingestion of a foreign object. The limits vary in accordance with the part of the blade where the damage is located. The portions of the blade which have higher stresses may not be cut as deeply as the portions subjected to lower stresses

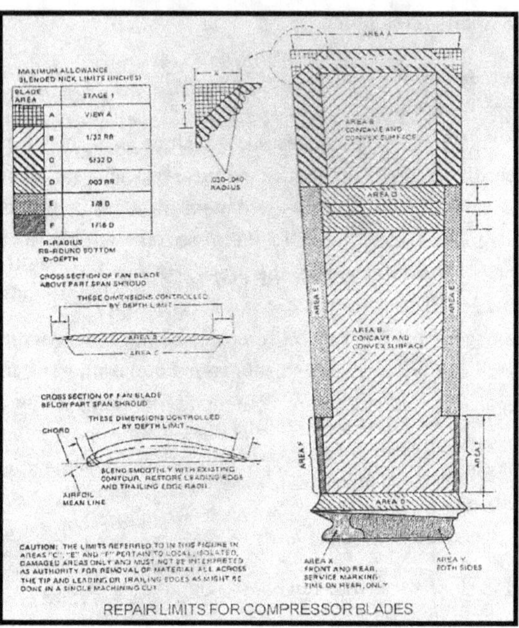

during operation. During blade repairs, care must be taken to maintain the original profile of the blade within reasonable limits.

The foregoing examples of blade repair are provided for information only, to illustrate typical practices. For a specific engine, the appropriate specifications given in the maintenance manual must be used.

Turbine Nozzles and Vanes

First-stage high-pressure turbine nozzles and vanes are "Subjected to the highest temperatures during operation since they are exposed to the gases as they exit from the combustion chamber. The high temperatures lead to expansion cracks, stress-rupture cracks, some burning, and other damage. Stress-rupture cracks usually appear along the leading edge of the blade. Manufacturers and operators have determined what amounts of damage are acceptable and will not affect the safe operation of the engine. The following list gives examples of serviceable limits for a high-b-mass engine the first-stage nozzle vanes.

Inspection (Borescope) and their serviceability limits

1. Axial cracks in the trailing edge (concave side only) in slots adjacent to trailing edge - Arty number 0.30 in. in length allowed provided they are 0.06 in. apart; or two per vane 0.80 in long, provided they are 0.30 apart; or one 1.50 in long with two 0.50 in long provided they are 0.30 in apart and do not extend forward of the leading edge gill holes.

2. Axial cracks in the leading edge - Any number 0.50 in long if separated by atleast 0.25 in or any number interconnecting the cooling holes provided the total length of interconnecting cracks does not exceed 0.60 in

3. Radial cracks in the concave surface between the inner and outer platforms -Any number 0.50 in long; or two per vane 0.80 in long provided they are at least 0.30 in apart

4. Radial cracks in the convex surfaces between the inner and outer platforms - one crack allowed 0.80 in long.

5. Blocked cooling air passages - Five nose holes and four gill holes in each row. A minimum separation of one open hole shall exist between blocked holes. Three trailing-edge slots may be blocked provided that blocked slots are not adjacent.

6. Nicks, scores and scratches - Any number, any length allowed if not over 0.03 in deep. Nicks up to 0.10 in deep and 0.25 in long allowed on airfoil hailing edge

7. Buckling or bowing of the trailing edge - Any number up to 0.30 in from original contour.

8. Axial cracks in concave surface - Two per vane extending aft from the first row of gill holes to (not enough) the slot in the trailing edge.

9. Axial cracks in convex surface - Two per vane between the midchord strut and the trailing edge, 0.25 in radially apart, total not to exceed 1.0 in; width not to exceed 0.50 in aft of the midchord strut. A maximum of three vanes per assembly not adjacent.

10. Burns in the trailing edge (loss of metal) - The total area removed from the trailing edge not to exceed 3.00 in Z [19.35 cm`] per assembly. Cumulative area is determined by summing individual vane area, radial height by axial length,

11. Burns and cracks on the convex and concave sides - Not to exceed an area of 1.50 in long and 1.0 in wide per vane. Maximum of four vanes per 900 arc.

12. Burns or spalling on vane leading edge (charred only, no holes through airfoil) -0.50 in diameter per vane, maximum of four vanes affected per 900 arc.

13. Craze cracking (craze cracking is defined cracking is defined as superficial surface cracks which have no visual width or depth) - any amount.

Turbine Blades

Serviceability limits for turbine blades are much more stringent than are those for nozzle vanes. This is particularly true for first-stage blades because of the high temperatures involved. The centrifugal stresses to which turbine blades are subjected require that the blades be free of cracks in any area and that no nicks or dents exist in the root area. .A limited number of small nicks and dents can be permitted in the areas of the blade away from the root area. No burning or distortion is permitted.

Other conditions to look for during inspection include blade creep, which is the permanent elongation of the turbine blades due to rotational forces, and untwist, which is a condition that results from. the gas path forces acting on the turbine blades. These forces tend to change the pitch of the blade, which generally decreases blade efficiency. A compressor - turbine wheel is shown in figure.

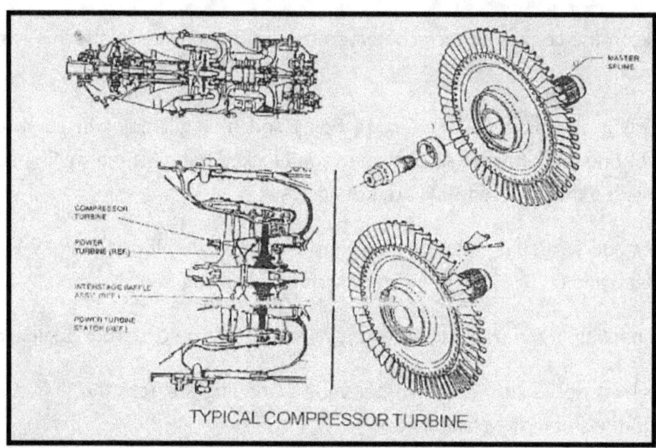

Repairs for Turbine Nozzles, Vanes, and Blades

When a borescope inspection reveals that there is damage or deterioration in the hot sections of the engine, the areas involved must be disassembled sufficiently to remove the defective parts. Parts requiring repair are replaced with new or reworked parts from the factory or an overhaul facility.

Replacement of turbine blades must be done with blades having the correct moment-weight designation to ensure that the turbine rotor will be in balance when assembled. The maintenance manual for each engine specifies the correct arrangement of blades according to their moment-weight markings.

www.ingramcontent.com/pod-product-compliance
Lightning Source LLC
Chambersburg PA
CBHW071952150726
47999CB00001B/420